Neuroscience and Christian Formation

Neuroscience and Christian Formation

Edited by

Mark A. Maddix
Point Loma Nazarene University

and

Dean G. Blevins
Nazarene Theological Seminary

Information Age Publishing, Inc.
Charlotte, North Carolina • www.infoagepub.com

Library of Congress Cataloging-in-Publication Data

CIP data for this book can be found on the Library of Congress website:
http://www.loc.gov/index.html

Paperback: 978-1-68123-673-5
Hardcover: 978-1-68123-674-2
E-Book: 978-1-68123-675-9

Copyright © 2016 IAP–Information Age Publishing, Inc.

All rights reserved. No part of this publication may be reproduced, stored in a retrieval system, or transmitted in any form or by any electronic or mechanical means, or by photocopying, microfilming, recording or otherwise without written permission from the publisher.

Printed in the United States of America

CONTENTS

Preface
  Dean G. Blevins and Mark A. Maddix vii

1. A Primer on Neuroscience
   Laura Barwegen 1

2. Technology and the Brain
   Dean Blevins 11

3. How Neuroscience Informs Teaching
   Glena Andrews 21

4. Neuroscience and Christian Formation
   Mark A. Maddix 33

5. Neuroplasticity and Spiritual Formation:
   Changing Brain Structure and Core Beliefs Through
   Mindfulness and Scripture Meditation/Reflection
   Karen Choi 45

6. We Were Made for This: Reflections on the Mirror Neuron
   System and Intercultural Christian Education
   Timothy Paul Westbrook 71

7. Christian Education as Embodied
   and Embedded Virtue Formation
   Brad Strawn and Warren Brown 87

8. Neuroscience and Christian Worship:
   Practices That Change the Brain
   Dean Blevins 99

9. Making Connections:
   Neurobiology and Developmental Theory
   Theresa A. O'Keefe *111*

10. Neurological Development in Early Young Adults
    and Their Implications for Christian Higher Education
    James R. Estep and John Trentham *121*

11. Changing Behavior and Renewing the Brain:
    Neuroscience and Spirituality
    Mark. A. Maddix and Glena Andrews *137*

12. Equipping Minds for Christian Education or Learning
    From Neuroscience for Christian Educators
    Carol T. Brown *153*

About the Editors *171*

PREFACE

Mark A. Maddix and Dean G. Blevins

Why a text on neuroscience and Christian formation? Simply put, we need one that represents the range of possible intersections for today and into the future. In recent years, neuroscience's various fields of study began to influence our understanding of the person, memory, learning, development, communal interaction, and practice of education. Professional organizations like the Religious Education Association (REA) and the Society of Professors of Christian Education (SPCE) found themselves dedicating entire annual meetings to the subject. Some of the chapters in this book began as presentations in those settings. As the research in neuroscience grows, the points of intersection between our understanding of the brain and our task in shaping spiritual lives seem inevitable. So, we offer this text as a group of scholars who may well feel more like "pioneers" exploring new territory than "settlers" residing in the well-worn conceptual neighborhoods of the faith formation of the past.

Readers may sense some redundancy of themes as each theorist seeks to unpack recent insights into neuroscientific studies of the mind, mirror neurons, theoretical systems, and the complexity of neuroanatomy that reveal a host of regions of the brain. In addition, core theological themes reappear including theological anthropology, divine causation, and the nature of the Christian community we know as the church. Hopefully the repetition of each of these themes creates an opportunity to allow certain sections to serve as an introduction to later chapters.

Readers will find an overall flow to the text. The book starts with four chapters that discuss the basic considerations for fashioning an understanding of how neuroscience influences Christian education and formation. Chapters five through eight then explore how neuroscience informs formational practices, from personal meditation, to intercultural encounter, to congregational formation and worship. The last four chapters explore how various aspects of neuroscience and Christian education intersect along developmental lines, addressing youth's decision making, young adult meaning and religious experience, and finally exploring children's ability to learn regardless of developmental ability. In addition, this book moves from more conceptual overviews at the beginning to more empirical studies late in the text.

Each chapter of this book can also be read and discussed individually. Each author has provided both discussion topics and suggestions for future reading within neuroscience and Christian education questions at the end of their chapter. Each chapter also represents varying views how neuroscience and faith formation intersect, some controversial, some competing against each other, and some views which may well pass away as research continues. Collectively our hope remains that the book will prompt ongoing investigation and reflection over this expanding scientific field and its influence on discipleship in and through our efforts in Christian formation. We do hope you engage in, struggle with, and enjoy this journey.

CHAPTER 1

AN INTRODUCTION TO NEUROSCIENCE AND CHRISTIAN FORMATION

Laura Barwegen
Wheaton College

INTRODUCTION

For some Christians, there remains a healthy skepticism in seeking correlations between scientific discoveries in neuroscience and the process of Christian spiritual formation of the soul. Sometimes it is articulated as the fear of reducing all of what it means to be human to the chemical exchanges and electrical impulses of our brains; other times this skepticism might be expressed as the fear of a demoralizing and deterministic model of humankind, or as atheist Richard Dawkins has put it, as viewing humans as very complicated and complex DNA replicators (Dawkins, 2006).

In his text *The Clockwork Image*, Donald MacKay coined the term "nothing-buttery" (1974). Nothing buttery, according to MacKay, "is characterized by the notion that by reducing any phenomenon to its components you not only explain it, but *explain it away*" (MacKay, 1974, p. 43). C. S. Lewis also warns against this tendency toward machine-mindedness in his essay on the *Abolition of Man*, citing the proclivity toward reducing humanity to "men without chests" (Lewis, 1944). In recent years, this fear has been articulated as ontological reductionism, the philosophical view of human persons that reduces them to nothing but the physical functioning of their neural mechanisms. As will be discussed in this book, Christian scholars like Nancey Murphy (2006) argue against this approach.

However, if we reduce our understanding of what it means to be human to our fears of nothing-buttery, what are we to make of advances in neuroscience and the documented implications in psychology, philosophy, theology, spiritual formation, biology, and other disciplines? For example, neurological research into Alzheimer's disease documents the effects of amyloid deposits throughout the brain, beginning with regions such as the hippocampus and temporal lobe (areas responsible for memory formation and recall), the prefrontal cortex (areas responsible for planning, working memory, and inhibition), and finally the parietal lobe (responsible for situating oneself in space) (Weaver, 2004). Matching these physical deficits with reports from the caregivers of Alzheimer's victims results in descriptions of a loss of a sense of self, or identity, a loss of feelings of the presence of the Holy Spirit, and a loss of a sense of a spiritual self. According to his research, Vogeley and colleagues document that deficits in aspects of identity—perceptions of our perspectives, thoughts, feelings, and actions—can be mapped onto neural dysfunction centered in the prefrontal cortex of the brain (Vogeley, Kurthen, Falkai, & Maier, 1999, pp. 343–363).

The purpose of this text is to identify correlations between neuroscience and Christian formation, written by a number of Christian authors and about a variety of different neurological processes—such as neuroplasticity and mirror neurons—so that we all might be better informed and able to draw conclusions about the extent to which Christians should consider these scientific advances. In order to better situate the information provided by each of these authors into a neurological framework and theological anthropology, some basic information is necessary about how neurons are structured and communicate; how change, or formation, is neurologically understood; and how to situate findings in neuroscience into the evangelical tent. Although this information may seem a little technical, it is important to have a reference to what is physically happening when authors throughout the remainder of the book refer to activity in the brain, or neurons "firing," or changes in neuronal connections.

How Neurons are Structured and How They Communicate

Neurons are the cells of the nervous system comprising, but not limited to, the central nervous system, the brain being the primary component.

We are born with approximately 100 billion neurons, and in the course of the first 2 years of our lives, this number is reduced to approximately 50 billion. Truth be told, this is good news! Neurons stimulated, and therefore used, by the experiences and stimuli provided in our early environments, begin connecting to other neurons to form neural networks. Neurons not stimulated, and therefore not used, by the experiences and stimuli provided in our early environments do not connect and, over time, will die and will not be replaced. A good case in point involves language development. In our first years of life, our auditory cortex seeks input with the neurons available to be used to hear specific sounds. Prevalent sounds within our early environments stimulate activity in neurons that have been intentionally designed by God to be activated by a specific pitch, tone, volume, and speed. If we do not hear specific sounds in our environment, neurons that have been created for the possibility of that sound die. We experience this as adults when we try to learn a new language. Even though we may become fluent, we can never replicate the sound of the language in the same way as a native speaker. We no longer possess the auditory structure to hear or replicate those sounds, so our brain makes the next best choices and utilizes neuronal columns in the auditory cortex that are "close enough."

In order for a neuron to be "activated," it needs to first receive stimuli from a source, and communicate this stimulation to another neuron. Neurons, like people, can do very little on their own; instead, they produce effects, such as movement, thought, emotion, and speech, when they act in concert with groups of neurons, called neural networks. The created structure of a neuron is designed for both incoming and outgoing information (See Figure 1.1).

Four segments comprise the basic structure of the neuron: the dendrites, the cell body, the axon, and the axon terminal. Each segment plays a role in moving stimuli, or information, from one neuron to the next.

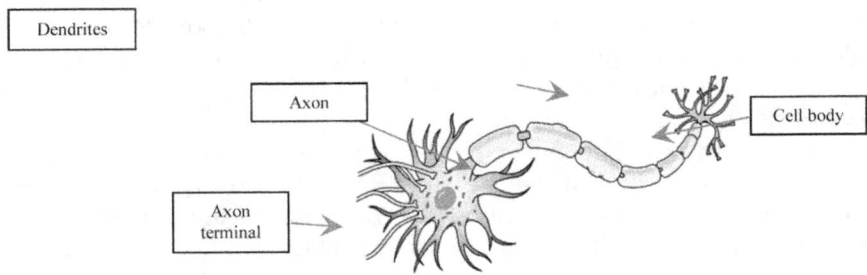

Figure 1.1.   Neuron.

Keep in mind that you have approximately 50 billion neurons in your brain, each with approximately hundreds or thousands of connections to other neurons. In order for a "connection" to be made, both a chemical exchange and an electrical impulse are necessary.

In its inactive state, called the "resting potential," the inside of the *cell body* remains electrically at odds with the outside of the cell body, since the inside of the cell is approximately -65 millivolts below the outside. This creates an electrical imbalance, with the positively charged sodium ions (Na+) pressing against the cell body to enter, but without open access. This is the condition of a cell that is not firing and not communicating with other neurons, called the "resting potential."

A second structure of the neuron is the dendrite. There are numerous dendrites that extend from the cell body and make connections with other neurons. Imagine your hand as a metaphor for a cell body and its dendrites. The palm represents the cell body and each of the fingers represents dendrites. Dendrites have ion channels, or gates, that remain closed to the positively charged sodium ions that exist in the cerebrospinal fluid surrounding the neurons. It is like putting your hand in water. The water presses to get into your body, but the skin prevents this from occurring. However, there gates do exist that the water *could* get in, but these gates remain closed. These ion channels, or gates, can be opened to allow entry of the sodium ions into the cell body, but they first have to be unlocked in order to open. Chemicals called neurotransmitters possess the key to unlocking these gates. A neurotransmitter attaches itself to the keyholes located on the dendrite and opens the gate, allowing the positively charged ions to rush into the cell body of the partnered neuron. Because these sodium ions are positively charged, they change the electrical charge within the cell body. When enough sodium ions have rushed in, and the electrical charge reaches approximately +70 millivolts, the receiving neuron, or "postsynaptic neuron," depolarizes and is ready to send what is called an "action potential." In other words, the neurons "fires." For a neuron, there is no weak or strong level of charging, nor does the strength of the charge diminish dependent upon the length of the neuron. And the charge travels away from the cell body, down the single *axon* that each neuron possesses. Dendrites move the electrical charge *toward* the cell body; the axon moves the electrical charge *away* from the cell body.

Each neuron has only one axon, but it has many dendrites. The purpose of the axon is to move the electrical charge through the neuron down toward the axon terminals. These terminals include the split-ends of the axon and contain vesicles, or small bubbles that house neurotransmitters, or the chemicals that will open the gates of other neurons, located on the dendrites. These vesicles rest at the ends of the axon terminals. When the electrical charge, sent from the cell body down the axon, arrives at the axon terminals, it pushes the vesicles to the edge of the terminal where they "pop," so to speak, releasing neurotransmitters into the space, or gap, that exists, be it ever so small, between the axon terminal of the sending neuron (called the "presynaptic neuron") and the dendrite of the receiving neuron ("postsynaptic neuron"). This gap, called a "synapse," defines an essential point where the communication from one neuron to the next exists. You may have heard the phrase, "he is not firing on all cylinders," or "not all of her synapses are firing," to mean that the person is not making mental connections. These phrases come from our understanding of the physiology of neuronal connections. One can visualize this similar to blowing bubbles; when the bubbles hit the concrete sidewalk, or the grass, they pop and would release anything that resided within them.

Therefore, at the neuronal level, communication occurs from one neuron to the next through an electrical charge that runs from the cell body down the single axon to the ends of the axon terminals and then through a chemical exchange from the release of neurotransmitters from the axon terminal into the synaptic gap. The neurotransmitters unlock the gate for the positively charged ions to rush into the postsynaptic neuron through chemically locked gates located on the dendrites.

HOW CHANGE, OR FORMATION, IS NEUROLOGICALLY UNDERSTOOD

As stated above, not much happens when one neuron fires, or even if two neurons fire. It takes a network of neurons firing together to enact movement, emotion, or thought. As neurons are continually utilized in connections with other neurons or outside stimuli, their axons are coated with a fatty substance, called myelin. This myelin sheath increases the speed of the electrical charge rushing down the axon, as well as protecting the axon. In addition to the myelin sheath, postsynaptic neurons regularly receiving neurotransmitters from presynaptic neurons internally increase the number of "gates" which can be opened by the neurotransmitters. These gates increase the openings by which the sodium ions can enter and alter the charge within the postsynaptic neuron, causing it fire more quickly. Researchers call these internally created gates NMDA and AMPA receptors. As one could imagine, by having more gates, by which positively charged sodium ions can enter, increases the speed by which the voltage of the cell body of the postsynaptic neuron depolarizes, or "fires." Eventually, and over time, these individual neurons become so intricately connected to other neurons that it may appear that they fire simultaneously, creating what is called a neural network.

The question remains, what initially causes these neural networks to become established? The answer is *experience* and *agency*. Individual experiences provide the stimulus for neuronal responses and interconnections in the brain. Repeated experiences—in action, thought, and emotion—contribute to neurological connections and neural network structuring, become myelinated and strengthened. Our own agency uses the strengthened, myelinated neural networks as the lenses by which we experience and interpret our world. Therefore, both experience and agency prove important in informing one another. Thus, "change" occurs through shifts in either experiences and/or agency.

Neurobiology, Christian Formation, and the Evangelical Tent

In her book *Bodies and Souls, or Spirited Bodies*, Nancey Murphy posits that "we don't have souls and we all get along perfectly fine" (2006, p. 4). She, and many other evangelical Christians, including most of the authors in this text, would identify themselves as "monists." Monists do not believe, based upon Scriptural and neurological evidences, that humans possess a separate substance called a soul (Brown & Strawn, 2012; Brown, Murphy, & Malony, 1998; Green, 2008). Instead, they believe the word "soul" identifies the *functionality* of our physical matter, namely the brain-body relationship. Other Christians find this position anathema. They believe Scripture clearly teaches about the existence of a soul, so any belief otherwise reflects the very antithesis of Christian faith, especially in light of an eschatological view of eternal life (Cooper, 1989; Evans, 1981; Moreland, 2014). These positions represent an ongoing tension throughout the history of our common Christian faith on a variety of theological points. However, we have also found approaches that evangelical Christians use to live within these tensions. Therefore, one must consider a variety of perspectives which people can take as they address current research in neurobiology; particularly how those views affect our view of human persons, or our theological anthropology, and our approach to spiritual formation. Our question entails asking which perspectives fall under the "evangelical tent," and which perspectives remain clearly outside of it. Figure 1.2 provides a visual concept.

As Figure 1.2 demonstrates, there are two distinct positions represented: one position, called *monism*, which claims that humans are made of one substance, and it is physical. A second position called *dualism*,

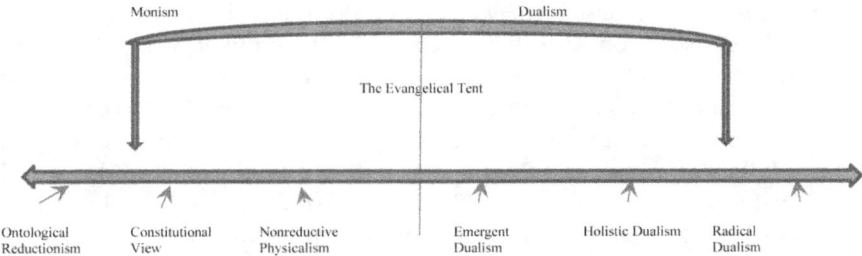

Figure 1.2. Anthropological views.

which claims that humans are comprised of two substances, a body and a soul.

Most of what many Christians fear about neurological research influencing our view of human persons is what is considered "ontological reductionism." This view represents the position which Donald MacKay described as "nothing buttery," and what many atheistic scientists, such as Richard Dawkins and Christopher Hitchens, espouse (Dawkins, 2006; Hitchens, 2007). This position, on the extreme end of Monism, states that everything we are as human persons (our ontology, our being) can be reduced to an explanation of our neurological functioning (reductionism). Holding this position precludes one from embracing the Christian faith, which prescribes a belief in a God who exists outside of oneself and acts within this world. God cannot be reduced to the neurological functioning of God's own creation. This position often remains historically rooted in an Aristotelian view of human persons: we are a composition of our physical constructions; and when we die, we cease to exist.

Radical dualism describes a different position on the opposite side of monism, yet one clearly outside of the evangelical tent. This position embraces a Platonic philosophy of human persons, one which fully separates the physical from the perfect forms. This position also proves untenable with the Christian faith, for it represents the heresy of *Gnosticism* (Quash & Ward, 2007). Gnosticism claims that the soul and the body remain completely separate substances, and the material must be rejected. The body serves as just a shell. Historical doctrines of Christianity—such as the doctrines of creation, resurrection, and human persons—speak against this heresy, and defend the value of the biological person.

However, Christians can find numerous positions which one could hold and still be within the evangelical tent. One might agree more strongly with one position over others, but each of these positions, briefly described below, situates one within the common elements of evangelical Christianity (Green, 2005):

- *Constitution View of Human Persons.* Presented by Kevin Corcoran, the constitution view firmly embraces a monistic view of human persons. Corcoran uses the example of a dollar bill. If one were to describe the composition of a dollar bill, one might cite paper and ink or other assorted chemicals. There is nothing different about a dollar bill from my copy paper. However, what gives the dollar bill value and meaning is the human person for whom it has great value. The constitution view of human persons states that humans are composed of matter and nothing else. However, what gives a person value and worth is God. Because God values human persons, they have worth and value. So much so that while we were yet sinners, Christ died for us (Romans 5:8).
- *Nonreductive physicalism.* Although this position also states that humans have only one substance, and it is physical, we are not reduced to only the physical functioning of our bodies. There a "nonreductive" component that allows for the existence of a higher being that supervenes upon us, although the specifics of how an immaterial substance, such as God, supervenes upon His material creation, remains a question. This is the position espoused by most authors in this text.
- *Emergent dualism.* William Hasker moves the conversation along the continuum into the realm of dualism, which argues that humans have both a body and a soul, as separate substances. For Hasker, the separate substance, which is the soul, emerges from the functioning of the physical body. Therefore, an emergent dualist might argue that the functionality of the brain, through the agency and experiences described above, creates this separate substance called the soul. This has significant implications for the ways that we live out our faith and are obedient to Scripture.
- *Holistic, or Minimal, Dualism.* John Cooper (1989) and Stephan Evans (1981) both argue for a form of holistic dualism, which posits that while we are mortal beings on this side of Heaven, we exist as physical beings, and all of the neurological research which correlates neurological functionality with spiritual formation would hold (holistic). However, at the point of death the souls separate from our physical bodies, thereby claiming that our souls are not separate, but are separable. For a holistic dualist there is still a respect for how neurobiological research informs the formation of our souls.

Both Christians and non-Christians may consider many more positions. The chapters throughout this text, written by brothers and sisters in

Christ, reflect a variety of ways that neurobiology informs for our spiritual formation, both as individuals and in communities.

DISCUSSION QUESTIONS

1. What are some initial fears you might have when considering the possibility of spiritual formation described in neurological terms? What does it mean to "take this too far"? On the other hand, what are the dangers of ignoring neuroscientific discoveries and how they influence thought, emotion, and identity?
2. Consider the possibility that the repeated experiences cause physical changes in the brain, and that the way your brain is physically wired causes you to interpret your experiences and respond to them in specific way. How does that awareness influence your choices? What does this say about the importance of the regular practice of spiritual disciplines?
3. Read Romans 12:1–2 and Ephesians 4:22–24. What might it mean if these Scriptures were viewed through a neurobiological lens? What if we were physically changed—the matter in our brains—by putting off the old self and putting on the new? How do you see discipleship differently because of this new understanding?
4. What is the difference between monism and dualism? Which position resonates most firmly with what you believe to be true?

REFERENCES

Brown, W. S., Murphy, N., & Malony, H. N. (Eds.). (1998). *Whatever happened to the soul?: Scientific and theological portraits of human nature.* Minneapolis, MN: Augsburg Fortress.

Brown, W. S., & Strawn, B. (2012). *The physical nature of Christian life: Neuroscience, psychology, and the Church.* Cambridge, England: Cambridge University Press.

Cooper, J. (1989). *Body, soul, and life everlasting: Biblical anthropology and the monism-dualism debate.* Grand Rapids, MI: Wm. B. Eerdmans.

Dawkins, R. (2006). *The god delusion.* New York, NY: Houghton-Mifflin.

Evans, C. S. (1981). Separable souls: A defense of 'minimal dualism.' *Southern Journal of Philosophy, 19,* 313–331.

Green, J. (2005). *In search of the soul: Perspectives on the mind-body problem* (2nd ed.). Eugene, OR: Wipf and Stock.

Green, J. (2008). *Body, soul, and human life: The nature of humanity in the Bible.* Grand Rapids, MI: Baker Academic.

Hitchens, C. (2007). *God is not great: How religion poisons everything.* New York, NY: Hachette Book Groups.

Lewis, C. S. (1944). *The abolition of man*. New York, NY: HarperCollins.
MacKay, D. M. (1974). *The clockwork image: A Christian perspective on science*. Downers Grove, IL; InterVarsity Press.
Moreland, J. P. (2014). *The soul: How we know it's real and why it matters*. Chicago, IL: Moody.
Murphy, N. (2006). *Bodies and souls, or spirited bodies?* Cambridge, England: Cambridge University Press.
Quash, B., & Ward, M. (2007). *Heresis and how to avoid them: Why it matters what Christians believe*. Peabody, MA: Hendrickson.
Vogeley, K., Kurthen, M., Falkai, P., & Maier, W. (1999). Essential functions of the human self model are implemented in the prefrontal cortex. *Consciousness and Cognition, 8*(3), 343–363.
Weaver, G. (2004). Embodied spirituality: Experiences of identity and spiritual suffering among persons with Alzheimer's dementia. In M. Jeeves (Ed.), *From cells to souls—and beyond: Changing portraits of human nature* (pp. 77–101). Grand Rapids, MI: Wm. B. Eerdmans.

SUGGESTIONS FOR FURTHER READING

Carter, R. (2014). *The human brain book*. New York, NY: DK.
Doidge, N. (2007). *The brain that changes itself: Stories from the frontiers of brain science*. New York, NY: Penguin Books.
Green, J. (2005). *In search of the soul: Perspectives on the mind-body problem* (2nd ed.). Eugene, OR: Wipf and Stock.
Moll, R. (2014). *What your body knows about God: How we are designed to connect, serve, and thrive*. Downers Grove, IL: InterVarsity Press.
Sacks, O. (1998). *The man who mistook his wife for a hat and other clinical tales*. New York, NY: Touchstone.
Schwartz, J., & Begley, S. (2003). *The mind and the brain: Neuroplasticity and the power of mental force*. New York, NY: HarperCollins.

CHAPTER 2

TECHNOLOGY AND THE BRAIN

Dean G. Blevins
Nazarene Theological Seminary

INTRODUCTION

The human brain may well represent the most complicated biological organism in creation. Filled with cells known as neurons, the brain houses more connections than there are stars in our universe. The depth and complexity of the human brain structure, let alone the intricate number of chemical interactions between those cells, stagger the imagination. No wonder Sir Thomas Willis' initial investigations of the human brain in the 17th century ushered in what Carl Zimmer (2004) calls the "neurocentric" era (pp. 6–7) when people turned from the heart to the brain as the seat of reason.

So why focus on technology? Why dedicate a chapter to massive machines, computer software, and microscopic exploration? Does this book represent an investigation into Christian formation and biology, or an explanation of physics and genetics? Also, how in the world does this relate to ministry?

Thanks to technology, researchers actually investigate human brains from normal people who remain neither incapacitated nor dead. In early studies of neuroscience, the only way to explore operations within the brain required an autopsy. Investigations concerning areas of hearing and speech often relied on abnormalities or afflictions (Seung, 2012, pp.

11–12). People suffering with either a brain incapacity or illness served as subjects of postmortem studies from the early age of neuroscience to the mid-20th century. Later, surgery patients yielded additional insights, but even those strategies left a number of key questions unexplored. The shift to noninvasive technology, and guided interventions under direct medical supervision, reveal new insights into brain functioning rather than brain abnormalities.

Today, we possess a number of approaches to gain a better understanding of the brain. Brain imaging techniques rely on different types of signals obtained from the brain. Researchers classify information according to the electrical, magnetic, optical, and chemical signals used to detect the brain activities. Unfortunately, our access still remains limited. As technology improves, so will our understanding. To help Christian ministers and educators appreciate the impact of technology, we begin by exploring how technology provides insight into the brain's synaptic firing, the blood flow that resources brain activity, and finally the future of research that employs computations, and even invasive incursions into brain activity, to deepen our understanding.

FIRING

While many forms of technology assist neuroscientists, most of the noninvasive approaches rely either on (synaptic) firing or (blood) flow. Brain cells, known as neurons, rely on a biochemical exchange of electrified particles across their synaptic appendages to form connections with other neurons. Brain cells transfer information along a neuron's synapse using an electrochemical exchange that strengthens the connections between those synaptic networks (Society for Neuroscience, 2012, pp. 7–12). One famous neuroscience maxim states, "neurons that fire together, wire together." In short, brain activity can be charted by the electrochemical "firings" of brain cells during mental activity.

Electroencephalogram technology, or EEG, identifies the electrical activity of the brain near the scalp through a set of sensors (often resembling a skull cap) to measure the brain's electric activity much like an EKG (electrocardiogram) can monitor heart activity. EEG measures at least four distinct rhythms of brain activity—delta, theta, alpha, and beta—in order of diminishing wavelength. This process both charts brain activity as a general activity, such as studies in the 1950s tracking the different stages of sleep, and identifies specific manipulations of arms and legs by paralyzed patients (Howard-Jones, 2008, p. 11). EEG technology can also measure cognitive activity, even when the learner may be unaware of the changes (Howard-Jones et al., 2010, p. 5). Researchers use EEG technology to track how children can discriminate between phonemes, and thus acquire languages, even when quite young (Posner, 2010, p. 36). Similarly, researchers monitor small variations in magnetic energy around active neurons (magnetoencephalography or MEG) to provide clues to localized neural activities.

EEG technology also provides an understanding of the structure, rather than the activity, of the brain when the initial "firing" comes from the technology rather than the neuron. For instance, researchers utilize X-ray and computer technology, known as computerized tomography or CT scans, to determine the size and mass of brains (Posner, 2010, p. 28). Overtime, CT scans provide the general structure of the brain, and also reveal tumors or other anomalies evident within the brain (Ward, 2006, p. 50).

In addition, large magnetic resonance imagery (MRI) machines provide another way of identifying the structure of the brain. The MRI represents one the most important innovations in brain research, whose developers received the 2003 Nobel Prize (Ward, 2006, p. 50). The MRI allows researchers to construct a three dimensional view of the brain. Protons, found in water in the brain, possess weak, random, magnetic fields. However, the MRI generates a magnetic field, defined in units called Teslas (T), that aligns some of the protons in a specific direction. To understand the order of magnetic intensity, most MRI machines generate strengths between 1.5 T to 3 T, while the earth's magnetic field remains around .0001 Tesla. Radio waves then knock the aligned protons to a perpendicular path (90 degrees). In this new state, the protons spin and generate a new signal until they gradually "return" to the original aligned state. Much like sonar, the echo change in location and return provides an image that emerges in small sections, known as voxels (three cubic mm). Depending on the magnetic strength of the MRI machine, and the rate of the "relaxation" of the protons, this activity also reveals different information concerning tissue in the brain (Ward, 2010, pp. 50–51).

MRI imaging provides a more in-depth view of the brain than CT scans, one that assists researchers' understanding of larger structural differences (such as between white and grey matter), and details concerning nuanced portions located deep within the brain. Such knowledge comes without surgery. In the case of some CT and general MRI scans, the "firings" do not rely on sudden shifts of normal brain activity, but they do yield considerable insight into brain structure. However, Christian educators remain more interested in how the brain functions, particularly as people learn. A specific form of MRI tracks brain functioning, one that relies less on direct EEG firings, but on indirect blood flow.

FLOW

Since cells create and send biochemical messages, they expend energy. Like all other organisms, the functioning brain's expenditure of energy resembles "work" or effort that requires the cells to replenish their oxygen (Ward, 2010, p. 53). The body delivers oxygen through hemoglobin or blood. So, researchers understand that where blood flows, brain cells "work." Researchers assume that increased blood flow corresponds to neural activity. If a person is engaged in a particular learning activity, blood moves to a specific portion of the brain. The challenge for a researcher rests with tracking the blood flow.

One way to track the flow of blood to an active brain region begins with the use of radioactive tracers injected in the blood stream. Using small portions of a radioactive isotope, the research uses technology similar to CT scans, yet designed to pick up the trace element using positron emission tomography (PET) scans. Most of the trace radioactive elements have very short shelf lives (ten minutes for some) so that the actual "window" for tracking the cognitive activity may prove quite small (Ward, 2010, p. 54). Still, the window yields remarkable results.

PET scans help neuroscientists identify regions of the brain heavily involved in basic tasks such as reading and listening, including subtasks associated with these activities. Researchers understand how different brain regions retain information, but others activate when a new pattern of information requires the brain to associate new material with older, established information (Posner, 2010, pp. 28–29). As Posner notes, "these findings supported the notion that mental operations occur in separate brain areas and showed how quickly these activations could be changed by practice" (p. 30). PET scans provide a foundation for educators and ministers to understand the brain's adaptability as well as speed of learning.

While PET scans prove important in capturing brain activity, they do appear problematic since the scans require the use of a radioactive isotope. A separate technology allows a less invasive approach to tracking activity in the brain. Neuroscientists utilize MRI technology to track functional activity (fMRI) as people engage in certain mental assignments. The process of tracking blood flow in the brain (and thereby identifying brain activity) often goes by the acronym BOLD, for blood-oxygen-level dependent. Granted, neural activity occurs throughout the brain at all times, if only to keep this organism alive. Yet scientists assume that a concentration of blood flows to places where the mental activity increases to accomplish cognitive tasks so that a concentration of voxel-sized activity provides evidence of the geographic location within the brain where said activity occurs.

Often the general public sees the end product of this research through vivid images of brightly colored activity superimposed on brains. In truth, fMRI technology actually identifies activity that resembles points of light on a radar screen. Mathematical computations then classify the intensity and concentration of the protons, often eliminating "outliers" that do not fit the overall activity. Computer software then converts these crude charts of activity into a smoother, more visually aesthetic, view superimposed on correlated scans of the same geographic region. While visually impressive, the real strength of fMRI technology rests with the ability to locate activity deep within the brain by tracking blood flow to specific areas.

Participants remain fairly immobile in large MRI machines, often engaged in mundane mental activities based either or responding to video imagery or repeating seemingly insignificant cognitive exercises. The physical limitations of many exercises, as well as the reliance upon computer software that often masks the imprecision of certain results, create suspicion that many reports overreach their stated conclusions (Satel & Lilienfeld, 2013). Truthfully, even tracking blood flow based on a cubic millimeter voxel remains fairly imprecise. Neurons, almost by definition, represent the most densely packed interconnected cells in the human body. Researchers note one voxel contains approximately 630,000 neurons with the potential for multiple synaptic connections (Aguirre, 2015). As such, the location of blood flow activity offers only general locations of brain activity, much like the maps of explorers like Christopher Columbus, in charting the general "geography" of the "new world" of the United States and the rest of North America. Drawing direct correlation, much less causation, between certain mental exercises and neural pathways seems fairly broad in generalization.

Nevertheless, researchers designing highly specialized, yet replicable, experiments prove quite capable of using fMRI results to infer reasonable observations about the brain and mental activity. Researchers themselves recognize the fragility of certain studies when popularized, or challenged through media sources (van Atteveldt et al., 2014). Instead, researchers often combine multiple studies before making generalizations concerning learning, behavior, or development (Spear, 2009). Neuroimaging studies help scientists and educators better understand a number of key mental tasks, from perception, recognition, and emotion, to calculation, to spatial navigation, to reading, remembering, and reward (Posner, 2010, p. 32). The range of cognitive tasks explored through neuroimaging reinforces the need for noninvasive strategies that allow researchers to continue exploring the brain.

As noted, outside of the use of radioactive isotopes, the advent of noninvasive technologies aid neuroscience researchers in their search to track cognitive processes in the brain. However, researchers also continue to use controversial, sometimes invasive, procedures that can change brain functioning. For instance, neuroscientists can employ magnetic waves to disrupt neural processes, known as TMS for transcranial magnetic stimulation (Patoine, 2010). Doctors often use this treatment for stroke victims. While noninvasive, since the stimulation does not breach the skin, inducing brain currents through TMS can prove controversial. Michael Persinger uses a similar process near the temporal lobes to create a sense of sensory displacement that he likens to religious experience. Persinger's conclusions, while intriguing, remain elusive at best (Hagerty 2009).

Other interventions do employ electrodes invasively deep within the brain of people suffering from moderate to severe neurological disabilities. Deep Brain Stimulation (DBS) represents a growing interest in using technology to suppress abnormal functioning or regulate current mental activities (Patoine, 2010; Talan, 2009). The use of intervention like DBS does create ethical concerns regarding its practice, similar to the psychosurgical lobotomies of a previous generation (Gray Matters, 2014, pp. 9–10). Nevertheless, both TMS and DBS point to a future that will include invasive procedures that may prove helpful as well.

FUTURE

Perhaps two major horizons appear in the future of technology and neuroscience. The first situates the smallest of synaptic connections and combines research with large scale computer projects. The second horizon incorporates even smaller technology which opens possibilities for future experiments and explorations within the brain.

To this point, researchers rely on technology primarily in understanding the geography of the brain. While intricate in nature, most studies explore larger physical sections of brain based on basic anatomy, particularly the frontal cortex (for rational judgment), the limbic or interior portions of the brain above the brainstem (for emotion and memory) and regions either to the back of the brain (for motor skills) or close to either of the temporal lobes for specialized activities like speaking or hearing. While intricate in nature, it appears easier to "map" brain activity primarily based on brain anatomy, seeing certain regions of the brain taking on "modular" mental activities. While crude (more neuroscientists actually believe in more intricate connections across the brain), this modular approach helps researchers locate and understand certain brain functions and relate them to learning tasks.

In the future, researchers may actually focus on the smallest point of interactivity between neurons, the synaptic exchanges or connections. Future studies may begin elsewhere since many of the brain's neuronal connections remain tightly interwoven, much like the interior of a baseball, so that pathways "weave" across the brain. Sebastian Seung represents a new generation of researchers that believe that the brain might be best described through the connections between and among these neural pathways. Seung envisions a comprehensive program, one that includes more careful dissection of existing brain material and one that begins exploring the synaptic connections across neurons. These connections, when combined into a regional or total map of activity, constitute a person's neuron's connectome. Replicating other studies of pathology, treatment, and learning, at the connectome level, Seung believes researchers might ultimately remap larger regions of the brain (Seung, 2012).

Seung's vision resonates with a version of computational neuroscience, which uses computers to replicate information processing similar to that of neurons. Using computer based models, combined with similar teaching software programs, may provide a "virtual" platform for anticipating connectomes in the brain (Thompson & Laurillard, 2014). The intersection between detailed study of neuronal pathways, and the computational gathering of insights into larger databases for computer modeling, represents the next major phase of research in Europe (Human Brain Project https://www.humanbrainproject.eu/) and in the United States (BRAIN initiative http://www.braininitiative.nih.gov/) reminiscent of the Human Genome Project Francis Colins (Sukel, 2015). As new insights emerge from these long-term neurotechnology projects, educators should possess a more nuanced understanding of the brain and learning.

A final technological advance may assist researchers to explore brain functioning in a manner that proves less invasive due to size. The advent of nanotechnology, synthetic devices designed at a microscopic level, afford an opportunity to create devices that can enter into the brain and deliver concentrated medications to a given region with little or no adverse reaction (Cetin, Salih, & Feyza, 2012; Hasan & Jagannadh, 2012). In the future, similar devices might be designed to cross the blood-brain-barrier, to record information in a specific region or subregion of the brain, or to therapeutically alter DNA of specific neurons and connectomes (Silva, 2006). For older ministers who remember the 1966 movie, *Fantastic Voyage*, this strategy may resemble more science fiction than substantive theory. Still, centers already exist to foster collaborative research not only in neuroscience but also nanotechnology and other forms of physics (Kavali Knowledge Connection, 2015).

As technology advances, neuroscientists and researchers possess an opportunity to continue to refine their knowledge of the brain. In the near future, general reports will rely primarily on research studying the brain's firing (via EEG) and flow (PET and fMRI) using conventional research. Still, the fact that researchers can systematically engage cognitive research without harming their subjects allows ministers and educators opportunities to read about the brain's processing of experience and learning. Recognizing that such reporting requires research that can be replicated and verified should caution Christian educators and ministers in their reception of simplistic journalistic reports. In addition, any intervention within the brain requires careful, ethical, consideration concerning the patient and their rights (Gray Matters, 2015). The contributions of technology should assist Christian educators in understanding the role of the neural pathways in people's experience and cognition.

DISCUSSION QUESTIONS

1. Why is it important to be able to use noninvasive approaches to study the brain?
2. .If we can identify basic regions that influence how we process hearing and reading, how might that help determine which educational strategies work best?
3. If blood flow seems imprecise, why might it still be helpful in understanding learning processes?
4. How might understanding learning at the synaptic level, even though computer models of learning and teaching, assist us as Christian educators?
5. What are some of the ethical concerns around the use of direct interventions in the brain like deep brain stimulation?

REFERENCES

Aguirre, G. (2015). *Number of neurons in a voxel*. Retrieved from https://cfn.upenn.edu/aguirre/wiki/public:neurons_in_a_voxel

Cetin, M., Salih G., & Feyza A. (2012). Nanotechnology applications in neuroscience: Advances, opportunities and challenges. *Bulletin of Clinical Psychopharmacology, 22*(2), 115–120.

Gray matters, volume 1: Integrative approaches for neuroscience, ethics, and society (2014). *Presidential Commission for the Study of Bioethical Issues*. Washington, DC. Retrieved from http://www.bioethics.gov

Gray matters, volume 2: Topics at the intersection of neuroscience, ethics, and society (2015). *Presidential Commission for the Study of Bioethical Issues*. Washington, DC. Retrieved from http://www.bioethics.gov

Hagerty, B. B. (2009). *Fingerprints of God: The search for the science of spirituality*. New York, NY: Riverhead Books/Penguin.

Hasan, A., & Jagannadah, B. (2012). Nanotechnology in medicine: Fine cures for the future. *Journal of Medical & Allied Sciences*, *2*(2), 29–30.

Howard-Jones, P. (2008). Potential educational developments involving neuroscience that may arrive by 2025. *Beyond current horizons: Technology, children, schools and families*. Bristol, UK: Futurelab. Retrieved from http://www.nfer.ac.uk/futurelab/

Howard-Jones, P., Ott, M., van Leeuwen, T., & De Smedt, B. (2010). *Neuroscience and technology enhanced learning*. Bristol, UK: Futurelab. Retrieved from http://www.nfer.ac.uk/futurelab/

Kavali Knowledge Connection. (2015). *The Kavali foundation*. Retrieved from http://www.kavlifoundation.org/science-spotlights/knowledge-connection#.VoF82BorLOQ

Patoine, B. (January, 2010). Progress report 2010: Deep brain stimulation. *Dana Foundation Report*. Retrieved from http://www.dana.org/Publications/ReportDetails.aspx?id=44348#sthash.zy79ZnYz.dpuf

Posner, M. (2010). Neuroimaging tools and the evolution of educational neuroscience. In D. A. Sousa (Ed.), *Mind, brain, & education* (pp. 27–44). Bloomington, IN: Solution Tree Press.

Satel, S., & Lilienfeld, S. O. (2013) *Brainwashed: The seductive appeal of mindless neuroscience*. New York, NY: Basic Books.

Seung, S. (2012). *Connectome: How the brain's wiring makes us who we are*. New York, NY: Houghton Mifflin Harcourt.

Silva, G. A. (2006, January). Neuroscience nanotechnology: Progress, opportunities and challenges. *Nature Reviews Neuroscience*, *7*, 65–74. doi:10.1038/nrn1827

Society for Neuroscience. (2012). *Brain facts: A primer on the brain and nervous system*. Washington, DC: Society for Neuroscience. Retrieved from http://www.brainfacts.org/

Spear, L. (2009). *The behavioral neuroscience of adolescence*. New York, NY: W. W. Norton & Company.

Sukel, K. (2015, November 30). Big data and the brain: Peeking at the future of neuroscience. *News at the Dana Foundation*. Retrieved from http://www.dana.org/News/Big_Data_and_the_Brain___Peeking_at_the_Future_of_Neuroscience/#sthash.4t3CVrT1.dpuf

Talan, A. (2009). *Deep brain stimulation: A new treatment shows promise in the most difficult cases*. New York, NY: Dana Press

Thompson, M. C., & Laurillard, D. (2014). Computational modeling of learning and teaching. In D. Mareshcal, B. Butterword, & A. Tolmie (Eds.), *Educational Neuroscience* (pp. 46–76). New York, NY: Wiley.

van Atteveldt, N. M., van Aalderen-Smeets, S. I., Jacobi, C., & Ruigrok, N. (2014). Media reporting of neuroscience depends on timing, topic and newspaper type. *PLoS ONE*, *9*(8), e104780. doi:10.1371/journal.pone.0104780

Ward, J. (2006). *The student's guide to cognitive neuroscience*. New York, NY: Psychology Press.

Zimmer, C. (2004). *Soul made flesh: The discovery of the brain—And how it changed the world*. New York, NY: Free Press.

SUGGESTIONS FOR FURTHER READING

BrainFacts.org Available online (12/24/2015) www.brainfacts.org
Dana Foundation Available Online (12/24/2015) http://www.dana.org/
Howard-Jones, P., Ott, M., van Leeuwen, T. and De Smedt, B. (2010). *Neuroscience and Technology Enhanced Learning*. Bristol, UK: Futurelab. Retrieved from http://www.nfer.ac.uk/futurelab/
Seung, S. (2012). *Connectome: How the brain's wiring makes us who we are*. New York, NY: Houghton Mifflin Harcourt.
Society for Neuroscience. (2012). *Brain facts: A primer on the brain and nervous system*. Washington, DC: Society for Neuroscience. Retrieved from http://www.brainfacts.org/
Sousa, D. A. (2010) *Mind, brain, and education*. Bloomington, IN: Solution Tree Press.

CHAPTER 3

HOW NEUROSCIENCE INFORMS TEACHING

Glena Andrews
George Fox University

I never intended to teach. In fact I was determined *not* to teach. As a graduate student, though, I found myself facing a room full of undergraduate students with a textbook in hand. I consoled myself believing this would be good one-time experience for me. Seven years later, my first full-time career position was as a college professor. I was trained as a neuropsychologist. How did I end up in front of a classroom of college students?

Over the past 25 years of teaching, I have continually utilized my understanding of neuropsychology and neuroscience to inform my techniques within the classroom. No matter what topic I teach, my training, research, and experience in neuroscience and psychology provide important guidelines on how to organize information for efficient encoding, present material to sustain attention, develop evaluations for best recall, and motivate students to put forth the effort necessary to successfully learn.

APPROACHES TO TEACHING CHRISTIAN FORMATION

An intriguing question occurs whether the teaching techniques I use in a course on neuropsychology change for teaching Christian formation? What are the differences in the topics? What would students bring to these courses that might lead to various interest levels and motivational states? A level of interest for the topic represents a mitigating factor for both areas of study. Whether learning about the regions of the brain or the definitions of faith and discipleship, the student is faced with facts to learn. The greater the interest, the better the encoding and thus memory of the facts. For both topics, the more solid the foundation of content and preparation, the easier the laying down of expanded memory schemas. Both neuropsychology and spiritual formation can begin with concrete facts but must progress beyond due to the nature of the abstract concepts. Learning facts as a way to "indoctrinate" students about religious ideas falls far short of the goal of teaching discipleship. Further, the approach to a group of students who have never been exposed to the idea of discipleship or faith practices would be very different from the classroom of students who have a background in the study of Bible, Christian history, and human cognitive development. For the novice or experienced student, the goal is not informational only; it must be transformational as well. The question then is, how does learning change us?

Learning is change (Zull, 2002), even according to early psychologists, and the change occurs in the brain. Thus, teaching entails "the art of changing the brain" (Zull, 2002, p. 17). Zull suggests teaching begins with concrete experience (stimulus), sensory experience (the beginning of the memory system), and moves to reflective observation, abstract hypothesis (internal analysis of comparisons), and active evaluation of data gathered (determining the accuracy of the stimulus presented). Learning entails a journey into "unexplored territory" (Zull, 2011, p. 23) rather than a specific achievement with an end. Whether we are guiding a group of children or a class of seminary students, we lead people on a journey of "changing the brain," one that can result in an altered life. This altered life hopefully exhibits changed practices, as well as changed knowledge, for the learner.

Concrete and Abstract Ability

Few would argue that teaching faith development to a group of 10 year olds should differ from teaching the same concepts to a group of 20 year olds. However, the only difference often initiated rests in changing language. We use smaller, more concrete words with younger children. Yet, it is not just the language of the two groups that differs. Differences include the brain's ability to understand the concepts of faith; the brain's ability to manage the emotional elements associated with belief, faith, and fellowship; and the brain's ability to elicit behaviors that enable our faith to mature. The differences occur in part due to the capacity to learn and recall (schemas). However, differences also involve the development of our ability to think in abstract ways, prioritize, and think into the future (plan). Put directly, faith remains a difficult concept to teach and understand because of its *abstract* nature. Even adults may not be able to fully appreciate abstract concepts; concepts for which we have few concrete or tangible associations or definitions. Words such as God, belief, faith, and love might have linguistic icons associated with them (e.g., heart for love) but our brain must develop a way to remember, think about, and manage these types of terms. We tend to initially learn primarily using our left hemisphere and its ability for sequential, concrete, verbally based thinking, before we move to abstract thought

Children, ages 3–18, were asked to draw pictures of God (Newberg & Waldman, 2010). Themes emerged based on the age of the child and the religious affiliation of the parents. Children between 3 and 10 years of age drew faces and people when asked to draw God. As children aged, the faces changed to drawings of crosses, hearts, or other symbols. The oldest children used pictures of the sun or lights to represent God. This pattern proved consistent with children from Christian environments 90% of the time. For children from non-Christian environments, 80% continued to draw faces and people to represent God even by age 16 (see Helmut Hanisch, 1996). These results suggest that growing up in a Christian environment influences the development, and possibly maturity, of our concept of God.

Readers should not be surprised that children were able to draw their conceptualizations of God. When children and adults struggle to manage abstract concepts, many if not most, tend to visualize. Our goal in teaching Christian discipleship entails providing a person with the tools to develop a more mature relationship with God that will result in a transformed life. So what is the difference between teaching the concepts and equipping our students to "solve" the problem of knowing how to develop a deeper relationship with God?

There are obvious differences between "doing" and "teaching" (Lee & Ng, 2011). It is helpful to understand not only what processes are activated for the learner to *compute* and solve concrete tasks, but of equal importance is the understanding of the processes necessary in order for the learner to acquire an understanding of *how* to do the task. Different areas of our brain are activated when learning by rote or by memorizing concrete definitions (left angular gyrus) versus learning by strategy (precuneus; Delazer et al., 2005). For example, we can teach the definitions of God, love, faith, and discipline to students so they are able to answer content questions. Learning content using rote memorization utilizes more verbal processes than learning to solve a problem using a strategy (Lee & Ng, 2011). Therefore, when children draw or describe a face (seen daily) when beginning to understand God, the action reflects a basic, concrete, understanding of the idea of God. It is closely related to memorizing "the facts." Our more mature student may struggle to draw or describe God, providing more abstract drawings, not because she or he is less capable of understanding God. Instead the student possesses a point which is similar to learning the strategy or the internal manifestation of a more mature relationship with God, one less easily verbalized.

One might possibly say we tend to use more verbal processes (left hemisphere) for the place and sequences of worship. As we guide students or group members to "move into" worship, we may be asking them to switch from a verbal processing to a more right hemisphere, abstract technique. These differences require the ability to understand abstract concepts and connect cognition with emotion. This process of change in how our brain responds to various levels of formation in worship is supported by research (Posner & Badgaiyan, 1997). Specific areas of our brain become more active (left frontal, temporal, and parietal lobes) when people read something meaningful. When the content of the reading seems more spiritual in nature, one finds greater activity in an even more specific area of the brain (parietal, Ramachandran & Blakeslee, 1998), highlighting the importance in understanding that Christian formation is a process that involves time, development, and understanding. The human brain seems intended to experience religious experiences, but no one specific area of the brain produces a religious experience...or experience of God (Newberg & Waldman, 2010). These studies help us to understand that teaching concepts of God and spiritual formation, understanding the methods for leading worship during which we want people to deepen their relationship with God is not a one-size-fits-all solution. We start with memorization of concrete definitions. Visualizations and verbal explanations must occur at the beginning of learning, but we do not want to stay there if we desire people to understand and experience more mature spiritual relationships.

Emotions and Learning

As suggested above, Christian Formation courses possess the potential to go beyond the learning of facts. We might consider that, by its nature, an emotional component to spiritual formation can lead to greater virtue. Christians and atheists contemplate God with equal depth and sincerity (Newberg & Waldman, 2010). Students can study the discipline of Christian Formation for just the facts, much as a person could take a class in math or history. But, if Newberg's work proves accurate, even a person enrolled in a course on formation will have a brain response that can lead to feelings of longing for an emotional or spiritual connection.

Our nervous system influences learning and experiencing. Our "fight or flight" response (sympathetic nervous system) activates when our brain perceives something to be feared. But interestingly, our system does not differentiate fear from love. Our brain, our neocortex, *interprets* the physiological sensations. Thus the same stimulus might be interpreted as frightening to one person and as exciting to another. Our social and cultural experiences in life help to shape our responses (Sousa, 2010). For example, if a person has been abused, the idea of "love" will not necessarily elicit feelings of comfort, belongingness, and positive affect, but rather fear, disgust, or anger. When we teach about becoming closer to a God who is loving, people have a variety of responses based upon their level of brain development influenced by life experiences.

The brain is wired to evaluate environmental situations (Willis, 2010). When there is a suggestion of a threat such as embarrassment or punishment, the brain (reticular activating system, RAS) considers the perceived threat first at the expense of any content that might be delivered by the teacher or professor. The result is that very little is learned because the student's brain system remains on high alert. The brain increases blood flow to the ventral area (front, center) of the frontal lobes, and decreases flow to the dorsal (back) area of the frontal lobes (Drevets & Raichle, 1998). Stressed students prove less capable of learning. When a student underperforms, it may be that she or he is overstressed (Jensen, 2008). Occurring stress overwhelms the areas of the brain that begin the process of managing emotions, so the student's ability to focus and make appropriate decisions decreases (Jensen, 2008). Alert to change, the RAS focuses on understanding the reason for the change. Because each person's mind determines stressors, the list of possible threats seem endless. Adding a new experience or structure to the classroom will warrant attention from the brain. Studies addressing the RAS suggests several possibilities for improving learning, the suggestions include creating a safe environment, introducing change associated with joy, and including physical activity within the learning experience.

When the brain processes information (sensory data), it merges with emotion and cognition; which defines learning. Learning also decreases when there is a lack of emotion (Damasio, 1994). The limbic system and basal ganglia (center of the brain) activate as emotions are experienced. As emotions become involved, our attention functions can be enhanced and thus potentially increase learning and memory. If anxiety increases though, the system for learning can become dysfunctional; attention is not consistent, working memory is faulty, and memories that are encoded often are not cohesive causing recall to decrease. Our logical brain can set goals but our emotions provide the motivation to move toward those goals (Damasio, 1994). Our emotions inform us about the level of importance of the material. A lack of emotion or an overly emotional state can impede our ability to learn (Jensen, 2009).

Executive Functioning

Executive functioning occurs in the frontal lobes (Willis, 2010). The prefrontal cortex includes functions like judgment, organization, prioritizing, assessment of risk, creating problems solving, concept development, and critical analysis. When we focus on faith development and the formation of beliefs about God and our relationship to God, we are in the realm of some of the most difficult beliefs and attitudes to change, because they utilize executive functions. For example, Christians usually value being raised in a Christian home, Christian based education, and specific church gatherings. It is difficult for people to change denominations or worship styles once they have been "raised" in a specific type of environment. Sometimes we see children, teens, or adults rebel, resist, or reject the teachings of their culture. A behavioral change that we view as dramatic, such as leaving the church, is actually physiologically easier than a shift in attitude or modifying beliefs. The dramatic nature of rejecting and leaving "shouts" to our brain, and we follow the behavior with the belief that we don't affirm the teachings of our former church, thus lessening what is termed cognitive dissonance. This behavior and chain of beliefs can be beneficial in situations in which the environment has been abusive. In order to avoid cognitive dissonance, the ones who leave and reject the church convince themselves that they no longer believe at all.

Changing or challenging the beliefs formed as children without rejecting or leaving them completely is complex. Thus, when we are working with people within the faith community or in the classroom, we must be mindful of the functions of the brain as we work to move the congregation, our Sunday school class, or our small group into a deep, more meaningful walk with God. When we engage in spiritual practices of silence, focusing on God or Jesus Christ, scripture, or prayer, certain areas of our brain become activated while other areas become less activated. Becoming less activated does not mean less activities; it means the brain is modifying "normal" activity. This means managing impulses. In order for a person to be able to manage impulses, there must be an inhibition of normal neuronal activity. Thus an impulsive or hyperactive person is really acting in the default mode of the brain, especially the frontal lobe. It is only with development and maturing that we are able to modulate our impulses. Because our brain does not always function effectively, some people are not able to modulate impulses without the assistance of medication.

Self and Other

When we are engaged in certain meditative spiritual disciplines, we can experience a loss of self. Decreased activation in the parietal lobe (the area aligned with reading meaningful material) is equated with a loss of sense of self (Jeeves & Brown, 2009). But during prayer or reading, the opposite occurs, with an increase in the activity of the right parietal area and increased frontal lobe activity (Azari et al., 2001) suggesting that prayer increases one's sense of self. The areas of the brain activated during religious experiences are not exclusive to religious experiences, but it is probably our mind that provides the interpretation of the neural activity (Jeeves & Brown, 2009). Religion is not reduced to a specific circuitry in the brain as is spoken language, but is more involved including cultural and social phenomena (Jeeves & Brown, 2009). Although spiritual formation may at first glance appear to be centered inward, toward oneself, in order to become mature in our spiritual discipline our Christian formation, we must have the capacity for empathy, for a sense of the otherness.

Learning Interference

We live in an age of multitasking and technology. We do not learn as well when distracted by texting, email, surfing the Internet, and other forms of electronic interruptions (Carr, 2011). Behaviors like surfing the Internet leads to shorter attention spans and decreased ability to make connections between concepts (Carr, 2011).

I began teaching with overhead projectors and transparencies and have "progressed" to teaching with Powerpoints embedded with video clips and animations. As technology becomes increasingly available, the rate of my students' learning decreases in the amount and the depth. Rather than engaging in multiple levels of activity, including listening, analyzing, and taking personal notes (episodic memory), students tend to sit passively without taking many, if any, notes. Instead, they play on their computers or phones believing they can just soak up the information and recall it later. During exams, they provide some information but not as much as earlier classes, and the depth of their understanding seems significantly lower.

What do we need to know about people with shorter attention spans, those with less ability to make connections between emotions and their motivation to learn, and those experiencing difficulty as beliefs change in the context of our congregations? Should we make our worship more exciting (to increase interest) with faster changes (to keep their attention) and superficial content (to match their decreased depth)? We experimented with these changes over the past couple of decades, yet people still struggle to find a meaningful faith and a meaningful relationship with God. Neuroscientist Champion-Jones (2014) suggests most human brains resist changes in schemas from pastors well trained to spend time contemplating and studying biblical passages in order to create *specifically deductive sermons* as directives for living a Spirit-filled life. She suggests the human brain would rather enjoy the challenge of discovering meaning. Interestingly, most homiletics textbooks now include, if not fully advocate, *inductive* sermons that powerfully engage the person in such discovery of meaning.

CHANGING THE BRAIN

Can we change our brains? The short answer is "yes." Our brains change every time we learn something new and our neurons make new synaptic connections. Christian ministers should consider key elements at this point. As psychologists who study learning and memory report, our memory system begins with the ability to *orient*. Orienting describes a physiological state of our senses alerting to something in the environment. We hear a sound and the sensory nerves in our ears activate. If a person cannot orient due to the inability to experience sensory information, learning cannot happen. This phenomenon includes overloaded sensory systems.

After orienting comes our ability to *attend*. Inattention is often misunderstood. Orientation and attention remain primarily functions of the frontal lobe. Because this is the last area of the brain to develop, it is the reason for the immature behaviors of children. When we see a lack of motivation, impulsivity, poor ability to prioritize, and poor immediate memory, it is due to the inefficient functioning of the frontal lobe. This area of our brain continues to develop into our mid-20s. With discipline and practice the neuronal connections between the limbic system (emotion), temporal and parietal lobes assist in the development of the frontal lobe or what we might call *maturity*.

Our working memory is basically housed in the frontal lobe as well. Thus the first three essential areas of our memory system and our ability to learn are dependent on the functioning of our frontal lobe. From the working memory, new information is processed through the short-term memory system which is housed primarily in the hippocampus of the temporal lobe. The information in the hippocampus is also short (a few seconds), but longer than working memory. The short-term memory is finite in space. When full, either no new information is processed, or old information is lost. If the information is not processed or is pushed out due to the attempt to cram too much information through to our long-term memory, we will have incomplete memories to recall (see Atkinson & Shiffrin model, 1971).

The interesting element about our long-term memory is that sections of our memory are housed in various areas of the brain. Thus when we need to recall information, we must recreate the memory. This is one of the reasons that we can have very different recollection of an event than another person at the same event. When we encode pieces of content into an area of the brain, we encode with sensory information as well...the smell of the coffee house as we were studying, the sense of the chair we were sitting in when the lecture was happening, the feeling of the person next to me moving around during the presentation. We even encode our mood along with the content of the memory. Thus when try to recall the information, being in the same place, next to the same person, and being in the same mood is the more efficient way to recall the information. I tell my students to study for an exam in the room where we were learning the information, and am as caffeinated when they are trying to recall the information as when they were trying to encode the information!

One of the methods often used for faith development has to do with behaviors. We encourage people to read their Bible daily, to pray, to use various methods for meditation including silence and solitude, repeated prayers or mantras, kneeling positions, or focusing on a song or pictures as they think about their relationship to God. Research in learning and memory inform us that to change our behavior and form a "new habit," we need to engage in the behavior at least twenty one days consecutively. That, in and of itself, proves hard to do. In my Psychology of Learning course I required the students to select a behavior they wanted to change and develop a plan for that change. To develop the behavior change plan, students needed to keep a record of the amount of time spent on the activity, such as devotions, for the two weeks prior to starting the new plan. The student would determine how many times a week and how many minutes a day would be spent in a devotional activity. Most of the time students would decide to read the Bible or a devotion book for 10 minutes at least 4 times a week. Finally, students were required to identify a reward for engaging in the new behavior.

We make errors in thinking when we believe that a person can engage in a new behavior for the intrinsic reward. Motivators vary between individuals and change across a lifetime. A young child can be encouraged to be quiet in church with a picture book or a piece of candy. Teenagers can be motivated to clean their room with the reward of permission to spend time with friends, on the computer, or with a favorite activity. The problem with adults, our wealthy world, is that we tend to allow ourselves all the rewards we want. My students often struggled to find a reward powerful enough to encourage them to spend 10 minutes a day in an activity that was not immediately intrinsically rewarding. If they decided they would not allow themselves to spend time with friends until their devotions were completed, it would often occur that the friends were very convincing and they would "promise" themselves to do the devotions before they went to bed. My students believed that devotions were important to deepen their relationship with God, but struggled to fit the 10-minute routine into their day.

When we teach a topic such as Christian formation, part of the struggle entails motivating people to engage in behaviors and new habits for which there is very little initial reward. If Christians cannot spend 21 days changing a behavior, like reading, or time alone, how can they deepen their relationship with God sufficiently to develop a "craving" for the time spent with God? Developing relationships with a tangible other (a person) takes time, but we need the person there to motivate us. We tend to develop the deepest relationships with people in close proximity. We feel the closest to those people with whom we have encounters frequently. We tend to fall in love with those in close proximity. So to encourage people to develop a deep, meaningful relationship with an abstract Other (God) is a challenge. Even though my students selected a behavior to change, developed all the target behaviors, and selected the reward, the changing of their behavior proved a challenge given all the distractors and countermotivators that occurred in their daily lives. Change often requires community.

When people find themselves in dire straits they often engage in dramatic programs to change habits. People enter residential programs in which they attempt to change their lives. We see this behavior with substance abuse programs, weight reduction programs, eating disorders programs, and trauma treatment programs. The person's entire environment is changed so that behavior change can be supported 24/7. Even after engaging in these programs, people still need to continue with support groups, therapy, and purposeful daily activities to continue their pathway in this new life. But we seldom feel the critical nature of spiritual neglect. We may be nearing "spiritual death," but do not sense that we are in "dire straits." If this is the case, what would motivate us toward spiritual rehabilitation? At the very least, we can utilize what we know about the brain to aid in teaching and discipleship.

Neuroscience informs our teaching of Christian formation through our understanding of brain development, the influence of environment on neuron synapses, and understanding of learning and memory and the process for behavior change. The good news is that the human brain can undergo change throughout our entire life. We can learn new thoughts, new facts, and engage in new behaviors throughout our lifetime. The good news is that a relationship with God can eventually become one of our most rewarding relationships. As with all development processes, those of us who are teaching on spirituality must also be informed by science in how to appropriately deliver information about faith, God, love, and a relationship with the Divine. We must attend to environmental and developmental factors as we engage in training people in ways that have as their telos a deepening relationship with an abstract Other. We need the research to inform us of methods for change in order to support and guide those who desire a life in which they act in ways that will lead to a deeper walk of faith.

DISCUSSION QUESTIONS

1. With this understanding of brain involvement and development, how might our instruction in spiritual formation differ across the lifespan?
2. What implications might there be for guiding people into a deeper relationship with God with our current understanding of the parietal lobe being activated both for our sense of self and our experience of meaningful reading and teaching?
3. Does worship need to be "emotional" to be meaningful? How might you address and support this question given the information in this chapter?

REFERENCES

Azari, N. P., Janpeter, N., Wunderlich, G., Niedeggen, M., Harald, H., Tellmann, L., ... R. J. Seitz (2001). Neural correlates of religious experience. *European Journal of Neuroscience, 13,* 649–1652.

Carr, N. (2011). *The shallows: What the Internet is doing to our brain.* New York, NY: W. W. Norton & Co.

Champion-Jones, P. (2014). *Brain-based worship: Remembering the mind-body connection.* Retrieved from http://digitalcommons.goergefox.edu/dmin/72

Damasio A. R. (1994). *Descartes' error: Emotion, reason, and the human brain.* New York, NY: G.P. Putnam.

Delazer, M., Ischebeck, A., Domahs, F., Zamarian, L., Koppelstaetter, F., & Seidentopf, C. S. (2005). Learning by strategies and learning by drill: Evidence from an FMRI study. *Neuroimage, 25,* 838–849.

Drevets, W. C., & Raichle, M. E. (1998). Reciprocal suppression of regional cerebral blood flow during emotional versus higher cognitive processes: Implications for interactions between emotion and cognition. *Cognitive & Emotion, 12*, 353–385.

Hanisch, H. (1996). *The graphic development of the God picture with children and young people: An empirical comparative investigation with religious and non-religiously educating at the ages of 7–16*. Stuttgart and Leipzip, Germany: University of Leipzig. (Translation from German).

Jeeves, M., & Brown, W. S. (2009). *Neuroscience psychology and religion*. West Conshohocken, PA: Templeton Foundation Press.

Jensen, E. (2008). *Brain-based learning: The new paradigm of teaching* (2nd ed). Thousand Oaks, CA: Corwin.

Lee, K., & Ng, S. F. (2011). Neuroscience and the teaching of mathematics. *Educational Philosophy and Theory, 43*(1), 81–86. doi:10,1111./j.1469-5812.2010.007.x

Newberg, A., & Waldman, M. R. (2010). *How God changes your brain*. New York, NY: Ballantine Books.

Posner, M., & Badgaiyan, R. (1997). Time course of cortical activations in implicit and explicit recall. *Journal of Neuroscience, 17*, 4904–4913.

Ramachandran, V. S., & Blakeslee, S. (1998). *Phantoms in the brain: Probing the mysteries of the human mind*. New York, NY: William Morrow.

Sousa, D. A. (Ed.). (2010). *Mind, brain and education: Neuroscience implications for the classroom*. Bloomington, IN: Solution Tree Press.

Willis, J. (2010). The current impact of neuroscience on teaching and learning. In D. A. Sousa (Ed.), *Mind, brain and education: Neuroscience implications for the classroom* (pp. 27–44). Bloomington, IN: Solution Tree Press.

Zull, J. E. (2002). *The art of changing the brain*. Sterling, VA: Stylus.

Zull, J. E. (2011). *From brain to mind: Using neuroscience to guide change in education*. Sterling, VA: Stylus.

SUGGESTIONS FOR FURTHER READING

Brown, W. (1998). Cognitive contributions to soul. In W. Brown, N. Murphy, & H. N. Malony (Eds.), *Whatever happened to the soul? Scientific and theological portraits of human nature* (pp. 99–126). Minneapolis, MN: Fortress Press.

Carr, N. (2011). *The shallows: What the Internet is doing to our brain*. New York, NY: W.W. Norton & Co.

Champion-Jones, P. (2014). *Brain-based worship: Remembering the mind-body connection*. Retrieved from http://digitalcommons.goergefox.edu/dmin/72

Demazio, A. (1994). *Descartes' error: Emotion, reason, and the human brain*. New York, NY: Putman and Sons.

Jeeves, M., & Brown, W. S. (2009). *Neuroscience, psychology, and religion*. West Conshohocken, PA: Templeton Foundation Press.

Sousa, D. A. (Ed.). (2010). *Mind, brain and education: Neuroscience implications for the classroom*. Bloomington, IN: Solution Tree Press.

Zull, J. E. (2011). *From brain to mind: Using neuroscience to guide change in education*. Sterling, VA: Stylus.

CHAPTER 4

NEUROSCIENCE AND CHRISTIAN FORMATION

Mark A. Maddix
Northwest Nazarene University

INTRODUCTION

The field of neuroscience and religion continues to expand as researchers seek to answer the relationship between the brain and theology, and more broadly between the mind and religion (Newberg, 2009, p. 1). Researchers seek to answer some fundamental questions such as "Where is the soul?" "Is there a 'God spot' in the brain?" "Is there a specific module of the brain that mediates religious experience?" "Do humans have free will or does the brain determine our behavior?" These questions have particular significance for Christians concerned about formation and education. Some Christians might be skeptical of a scientific proof for religious experiences, but it could be that science can help Christians gain an understanding into Christian practices that help humans grow and develop.

The study of the mind and religious experience is not a new discussion. Both Eastern and Western religions have considered the relationship between the mind and human spirituality for thousands of years. Eastern religions have placed emphasis on the mind and human consciousness as a person strives for enlightenment. The Western religions place emphasis on the mind and the spirit or soul which makes connections to God. But more recently, with the growth of neuroscience, particular attention has been given to the mind/religion question. This new field of study is called *neurotheology*. Neurotheology is a field of study that seeks to draw conclusions about the truth of religious claims from the study of biological brain events. In other words, it is the investigation between the brain and theology.

The fundamental goals of neurotheology as a field of study include the following four goals:

1. To improve our understanding of the human mind and brain
2. To improve our understanding of religion and theology
3. To improve the human condition, particularly in the context of health and well-being
4. To improve the human condition, particularly in the context of religion and spirituality (Newberg, 2009, p. 18).

These goals provide Christian educators with the knowledge that theology has something to gain from through the interaction with the cognitive neuroscientific research, and the study of neurotheology can improve understandings of religious experience. Neurotheology can provide Christian educators with knowledge about the impact of particular spiritual practices on health and well-being.

NEUROSCIENCE AND HUMAN ANTHROPOLOGY

In a book about neuroscience and Christian formation, the foundational question to be addressed is related to our understanding of the body/soul dualism. Dualism goes back as far as Plato's views of *forms* and *ideas*. Plato's forms existed externally in a transcendent realm and served as something like blueprints for material entities. Aristotle rejected Plato's view of absolute forms and believed that material substances were inherent in material things that they formed. It is the form that gives a thing its operative and directs its development. The Platonic understanding was enhanced and Christianized in the early Church history of Saint Augustine, and later asserted by René Descartes during the Enlightenment. Descartes refuted Aristotle's view and returned to a radical dualism of mind and body that reflected Plato's view. Descartes held to Cartesian dualism which states that there are two kinds of foundation, mental and body. His philosophy states that the mental cannot exist outside the body, and the body cannot think. He believed that humans were different from animals in having a rational soul that was immaterial and interacted with the physical boy through the pineal gland (Jeeves & Brown, 2009, p. 109). Descartes view still impacts many scientists and theologians' view of human nature.

Even though most neuroscientists are in favor of the physical embodiment of the mind, scientists remain on both sides of the argument. This division is also reflected among Christian theologians and philosophers (Murphy, 2006). It raises questions about the existence of the soul, free will, and the afterlife. Neuroscientists and theologians who reject Cartesian dualism are often viewed as physicalist, who denies the existence of the soul and free will. This is a valid critique because, when viewing religious experiences, it can lead to reductionism, in which human experience is nothing but a movement of molecules or chemicals. Also, the question of human free will is related to humans being created in the "image and likeness of God." The *imago dei* implies that humans have moral agency and the capacity to use reason, which is what separates humans from animals.

All of these questions are not answered in this chapter since the primary scope is to give an overview of the issues at hand. The neuroscientific and biblical/theological understanding of the human nature is explored below. These findings provide an important theoretical foundation for Christian formation.

Nonreductive Physicalism and Dual-Aspect Monism

Many neuroscientists are concerned about the Cartesian dualism that leads to a bifurcation of the body and the mind. These neuroscientists argue that a nonmaterial Cartesian entity should not be separated from the body. They reject a dualistic view of the human person and argue that the continual experiences of behavior and environmental societal feedback that the mind becomes formed as a functional aspect of our brain and body (Jeeves & Brown, 2009). They hold that both dualist and reductive materialist view of human nature are undesirable on philosophical and theological grounds (Markham, 2007). The alternative is nonreductive physicalism, or Emergent, or dual-aspect monism which asserts that humans are entirely physical beings and all causes of human behavior are simple chemistry and physics. Nonreductive physicalism suggests this rejection of reducing all life to its simplest parts. Many causes in life lie in the emergent properties of the whole person. Thus, dual-aspect monism singles out neither the physical nor the mental aspects of the whole of our mysterious nature, but says both aspects are necessary to do justice to reality (Jeeves & Brown, 2009, p. 131).

For example, Jeeves and Brown (2009) illustrate that the relationship between brain activity and religious experiences could be disconcerting to some religious persons, especially Christians. It could impact how Christians interpret religious experience and personal faith conversions. Neuroscientists have presented data showing that many of the faculties, such as religious experience, once attributed to the mind or soul can now be explained as complex functions of the human brain. In other words, religious experiences were once exclusively dealt with in terms of the soul but are now being investigated through brain science. Such studies have also shown that there are no connections between religious experiences and specific parts of the brain. In other words, you cannot locate a particular religious experience in one part of the brain (Newberg, 2009). Some Christian neuroscientists argue that these religious experiences and brain anomalies point to our "creatureliness" and the embodiment of our mental and religious experiences. They say nothing with respect to whether these experiences are linked to a reality outside of ourselves (Jeeves & Brown, 2009, p. 133). These studies demonstrate that our most meaningful and deeply personal experiences emerge from, and take place within, our physical selves (Jeeves & Brown, 2009, p. 133). This is why people in a religious context are far more likely to interpret a brain event as a "spiritual" experience. Many Christians believe that religion and spirituality are engendered by our embeddedness in Christian community and all activities that entails, and not inner properties of the brain (Jeeves & Brown, 2009, p. 134).

Biblical/Theological Considerations

A broader exploration of the biblical and theological foundations of anthropology is needed to give justice to this argument, but the following provides a brief overview of the dualistic and monistic debate about human anthropology. There is much debate about human nature from theologians and biblical scholars. On the one side there are those that believe that the Bible supports a body-soul dualism, while on the other side there are those who believe it does not (Green, 2010). For those who embrace a dualist view, they are highly influenced by Greek and Hellenistic philosophy. They base their view that the New Testament writers, particularly Paul, were influenced by Greek thought. This view is most reflected in how many Christians view the relationship between the material and spiritual. They also believe that recent discoveries in science are threatening the authority of scripture by making claims against this dualist view. Many of them believe that what ultimately matters is the soul, because it is eternal and will continue in eternity. The body or material world is temporal and will ultimately decay and die. This view has a strong influence in the Church because of the heresy of Gnosticism, which was disputed by the early church fathers. This view has practical implications for Christian living because it focuses on the otherworldly orientation and inner experience of faith. Gnosticism spirituality continues to have a strong impact on Christian living.

On the other side of the argument, many biblical scholars believe in a monistic view of the human person that is reflected in the Hebrew understanding of the human person. The Hebrews and Jews, as reflected in the ancient Scriptures, believed that the human person included all aspects of the person. As Joel Green states, "that behind anthropological dualism stands not so much the Old and New Testament but earlier generations of scientists and their influence on the Christian theological tradition" (Green, 2010, p. 184). In other words, Green is arguing that Scripture witnesses to a monistic view of the human person, but scientists have influenced our understanding of the human person. Green continues by stating that,

> Scholars of the Hebrew Bible are almost unanimous in their conclusion that the portrait of human nature found in its pages is of a psychosomatic unity. Even if the New Testament evidence is more contested, it is nonetheless widely agreed that the Greco-Roman world of the first century AD was capable of supporting both monist and various dualist anthropologies, and that the New Testament, whose primary theological influence was the Septuagint (LXX), comes down especially on the monist side of the continuum. (Green, 2010, p. 185)

Green's analysis of the consistency among biblical scholars in support of a monistic view provides a foundation for the integration of theology and science. It gives evidence, to those who have concerns about the authority of Scripture, that the recent discoveries in neuroscience do not contradict the Biblical witness. In reality, Scripture and the findings of neuroscience are complementary in the understanding of the human nature.

SPIRITUALITY AND CHRISTIAN FORMATION

Where does spirituality take place? Neuroscientists assert that spirituality takes place within our physical bodies. Jeeves and Brown indicate that the spiritual dimensions of our lives are both *embodied* and *embedded*. *Embodied* means spirituality is embodied in human biological makeup. It includes the interdependent processes between the brain processes, cognitive processes, and behavior processes. Embodied beliefs and expectations are major factors in understanding some of the dimensions of spirituality. *Embedded* means spirituality and religion are embedded in Christian communities which include experiences, beliefs, and practices. These are not inner properties of the brain (Jeeves & Brown, 2009). In this regard, spirituality is firmly embedded within communities and cultures. It is through these embedded experiences and practices that sculpt the brain and bring about scientifically observed benefits of well-being. Therefore, the spiritual dimensions of our lives are embodied and embedded.

Given these realities it seems that a Christian view of human nature includes understanding the humans as physical beings (embodied) that are deeply nested in formative social networks (embedded). Since spirituality is embedded in physical, cultural, and social environments, it underscores its impact Christian formation. Therefore, Christian formation is developed, maintained, and manifested in communities, and is never separated from the physical realities of life.

Christian Community

The focus of this chapter is to explore how the Body of Christ (Christian community) can provide healing, renewal, and transformation through practices. The inherent struggle in this discussion is that much of Evangelical faith is expressed and experienced as a privatized faith, particularly in the Western world. Evangelical Christianity has given particular focus on individual salvation and personal experience over against a more embodied approach to Christian formation. A gospel of individual salvation is often preached and taught without a focus on the communal aspects of faith formation.

Also, the emphasis on Church growth and the proliferation of mega churches makes it very difficult for Christian formation. Many people are attracted to larger congregations because they can be autonomous, less involved, and emotionally detached from other people. In many ways they have bought into an entertainment approach to church which makes them passive participants. However, the more recent influence of the Emergent church and missional church movement has brought a fresh critique of this form of church. They value practices that are communal, not isolated or individualistic. Ecclesial practices engage people in communal life and mission together (Maddix & Akkerman, 2013). In many ways these approaches to church get much closer to the model of Church expressed in the book of Acts.

If persons are to become mature in Christian character (virtuous living) it requires them to be embedded within the Body of Christ. A relational approach to Christian community includes persons-in-relationship instead of individuals focused on more internal and privatized faith. Christian formation includes the reciprocal dynamics of relationship in community to ensure the healing, shaping, and forming of Christian character. The church is essential in healing and restoring embodied and embedded persons which require a rethinking of models of Christian formation that focuses on individualized persons seeking an encounter with God. The goal is to overcoming these privatized notions of Christian formation by giving emphasis to practices that form persons.

Much research has been done in the areas of social psychology, and that gives evidence of the importance human development that focuses on attachment theory and interpersonal relationship expressed through mentors or exemplars (imitation) (Armistead, Strawn, & Wright, 2010). In this view the process of change and formation of embodied persons are largely social and relational, and not based on individualized experiences. The thesis of this chapter is that Christian formation that is reflective of embedded and embodied persons includes Christian communities that give emphasis to exemplarity (imitation) and interpersonal relationships through small groups.

Exemplarity (Imitation)

Social scientists and neuroscientists provide ample support that emotions are embedded within the neural networks and entangled with embodied fields of energy. The work of René Girard (1987) argues that human being are fundamentally *mimetic* or *mimicry* since they tend to copy each other's behavior. In his view of mimesis it affects not only a person's outward behavior but also one's inner thoughts and feelings, one's desire and aversion (Maddix, 2015, p. 76). He argues that *mimicry* or imitation takes place through *mirror neurons* as feelings and intentions are enacted. Ron Schults argues that all of our social relations are shaped by our embeddedness within emotional systems that are transmitted across generations through patterns of mimicry within nuclear and extended families of origin (2013). He believes that the dynamic of human exemplarity are embedded with and have objective and potentially transformative effects in the real world (Schults, 2013). In other words, the transformation of humans is affected through and within the social dynamics of imitating and being imitated. Exemplarity is a generative and productive power for real transformation (Schults, 2013). Humans are always influencing each other's attitudes, beliefs, and behaviors through reciprocal imitation.

Neuroscientists also understand that the human brain has the capacity to rearrange and be changed through the process of neuroplasticity—the mechanism that allows change to occur in the brain (Bergley, 2007). It is through the repetition of practices, or in the case of mimicry, through the imitation of persons, that the brain can change and develop a great capacity for love and empathy.

The research on exemplarity is beneficial in understanding an embodied persons being shaped and formed through human relationship. The significant impact that exemplars play in Christian formation is substantial. It reminds local congregations that the interpersonal relationships established through mentoring and, in small groups, results in transformation and growth. Neuroscientists and social psychologists give us scientific evidence showing that persons embedded in relationships and community bring about maturation and growth.

Girard's theory has significant implications for Christian formation with the focus on sharing in Christ's life by mimetically participating in Christ's life, death, and resurrection. This is expressed as through participation in the Eucharist as eating and drinking the cup vividly reenacts the participation in Christ's divine-human subjectivity. It also includes our participation in baptism as identification, a portrayal of the union of Christ's death and resurrected life so that through identification with him we become "crucified" to spiritual bondage (Maddix, 2015, p. 81). Girard reminds Christians that the participation in these communal practices such as communion and baptism are more than mere symbolic actions, they provide a means of grace by which persons participate in the story and narrative of Christ. In other words, they enter into, and participate with Christ on an emotional level that changes the brain and helps them become imitators of Christ.

Small Groups

Another important practice that fosters transformation and growth of embodied and embedded persons is small groups. Small groups provide a context for deepening relationships and connectedness. They reflect the very nature of the relational Triune God. Humans are created as relational begins in need of acceptance, love, and care. Participation in small groups provides Christians a sense of well-being, connectedness, and ownership of the life and faith formation. Small groups can become another avenue where human attachment and imitation takes place. In the field of psychology attachment theory focuses on the need of a child to have secure relationships with adult caregivers. In the same way, adults need attachment to other adults through significant interpersonal relationships (Blevins & Maddix, 2010). Reciprocal imitation takes place as adults imitate practices, beliefs, and behaviors as they share life together. Adults in these groups can become attached through interpersonal relationship because of a sense of security and trust. In this way, these groups become a means of grace whereby embodied and embedded persons are nurtured and developed.

The benefit of small groups is that it moves beyond a privatized faith. Individuals are no longer isolated but become embedded in a physical community.

SUMMARY

As the field of neurotheology continues to expand in its investigation of religious experiences, it provides a context to understand the relationship between the brain and religion. Even though some Christians might be skeptical of the science of religious experience, the evidence of the impact of religious practices on the brain is well documented. Even though there is disagreement among neuroscientists and theologians about theological anthropology, the majority reject the philosophical and theological view of Cartesian dualism in favor of nonreductive physicalism as a basis to understanding human persons. A nonreductive physicalism gives expression to an understanding of humans as both embodied physical persons embedded in social contexts. This view rejects an individualistic view of human persons where Christian formation takes place in a privatized manner. The impact of this understanding provides Christian educators with an approach to Christian formation that includes a robust ecclesiology built around communal practices.

This ecclesiological shift requires many Christians to create a new imagination of the formative power of ecclesial communities. It also urges pastors, Christian educators, and laity to reconsider the formative role that social relationship play in Christian formation. They need to consider the influence of social psychology and neuroscience that underscores the important role exemplars. Since exemplarity is a generative and productive power for real transformation as Christians influence the attitudes, beliefs, and behaviors through reciprocal imitation. The impact of the repetition of practices, particularly communal practices such as small groups, participation in the sacrament, and service, by God's grace has the power to heal, renew, and change the brain.

DISCUSSION QUESTIONS

1. In what ways has Cartesian dualism impacted our understanding of human persons particularly as it relates to Christian formation?
2. What can we learn from research from social psychologists and neuroscientists regarding the importance of communal practices?

3. Why are many Christians reluctant to engage in smaller ecclesial communities or small groups? How can Christians overcome this barrier to Christian formation?
4. What is the impact of René Girard's view of *mimicry* as it relates to participation in the life, death, and resurrection of Christ?
5. What aspects of congregational life might your congregation need to develop in order to ensure that communal practices such as small groups, the sacraments, and service are included to foster faith formation?

REFERENCES

Armistead, M. K., Strawn, B. D., & Wright, R. W. (Eds.). (2010). *Wesleyan theology and social science: The divine dance of practical divinity and discovery.* Tyre, UK: Cambridge Scholars.

Bergley, S. (2007). *Train your mind, change your brain: How a new science reveals our extraordinary potential to transform ourselves.* New York, NY: Ballantine Books.

Blevins, D., & Maddix, M. A. (2010). *Discovering discipleship: Dynamics of Christian education.* Kansas City, MO: Beacon Hill Press.

Girard, R. (1987). *Things hidden since the foundation of the world* (S. Bann & M. Metters, Trans.). Stanford, CA: Stanford University Press.

Green, J. B. (2008). *Body, soul, and human life: The nature of humanity in the Bible.* Grand Rapids, MI: Baker Academic.

Green, J. B. (2010). Science, theology, and wesleyan. In M. K. Armistead, B. D. Strawn, & R. W. Wright, (Eds.), *Wesleyan theology and social science: The divine dance of practical divinity and discovery* (pp. 177–192). Tyre, UK: Cambridge Scholars.

Jeeves, M., & Brown, W. S. (2009). *Neuroscience, psychology and religion: Illusions, delusions, realities and human nature.* West Conshohocken, PA: Templeton Foundation Press.

Maddix, M. A. (2015). Moral exemplarity and relational atonement: Toward a Wesleyan approach to discipleship. *Wesley Theological Journal, 50*(1), 67–82.

Maddix, M. A., & Akkerman, J. R. (2013). *Missional discipleship: Partners in God's redemptive mission.* Kansas City, MO: Beacon Hill Press.

Markham, P. N. (2007). *Rewired: Exploring religious conversion.* Eugene, OR: Pickwick Publications.

Murphy, N. M. (2006). *Bodies and souls, or spirited bodies?* Cambridge, England: Cambridge University Press.

Newberg, A. B. (2009). *How God changes the brain: Breakthrough findings from a leading Neuroscientist.* New York, NY: Ballantine Books.

Newberg, A. B. (2010). *Principles of neurotheology.* Surrey, England: Ashgate.

Schults, F. L. (2013). Ethics, exemplarity, and atonement. In J. A. Van Slyke, G. Peterson, W. Brown, K. S. Reimer, & M. L. Spezio (Eds.), *Theology and the sci-*

ence of moral action: Virtue ethics, exemplarity, and cognitive science (pp. 170–172). New York, NY: Routledge.

SUGGESTIONS FOR FURTHER READING

Armistead, M. K., Strawn, B. D., & Wright, R. W. (Eds.). (2010). *Wesleyan theology and social science: The divine dance of practical divinity and discovery.* Tyre, UK: Cambridge Scholars.
Green, J. B. (2008). *Body, soul, and human life: The nature of humanity in the Bible.* Grand Rapids, MI: Baker Academic.
Markham, P. N. (2007). *Rewired: Exploring religious conversion.* Eugene, OR: Pickwick Publications.
Murphy, N. M. (2006). *Bodies and souls, or spirited bodies?* Cambridge, England: Cambridge University Press.
Van Slyke, J. A., Peterson, G., Brown, W., Reimer, K. S., & Spezio, M. L. (Eds.) (2013). *Theology and the science of moral action: Virtue ethics, exemplarity, and cognitive science.* New York, NY: Routledge.

CHAPTER 5

NEUROPLASTICITY AND SPIRITUAL FORMATION

Changing Brain Structure and Core Beliefs Through Mindfulness and Scripture Meditation/Reflection

Karen Choi
Presbyterian Theological Seminary in America

INTRODUCTION

As a former biology major prior to conversion with a special interest in the human brain, and now as a Christian educator, I remained fascinated by the influx of neuroscience research on mindfulness and mindfulness meditation in recent years. My interest in the human brain and my strong desire to understand the process of spiritual formation led me to explore how a Christian understanding of mindfulness might encourage change via the neuroplasticity of the brain.

What is neuroplasticity? Neuroplasticity refers to the ability of the brain to change "its structure, circuits, chemical composition, or functions in response to change in its environment" (Schwartz, & Gladding, 2012, p. 36). In more simple terms, neuroplasticity describes the way the brain's nerve cells, also known as neurons, change their structures and their connections with each other due to different stimuli. Before 1998, although the brains of infants and children were known to be plastic or malleable, researchers and practitioners widely accepted that neuronal connections in the adult brain were hardwired and immutable (Begley, 2006). However, with Eriksson and colleagues' (1998) discovery of neurogenesis (generation of new neurons) in the adult hippocampus, researchers now know that adult brains remain plastic and can continue to form new neural connections and generate new neurons in response to learning or training even into old age. Advances in neuroimaging techniques reveal how environmental factors, such as experience, training, and learning, can contribute to positive change, even at the neurobiological and *structural* levels throughout one's lifespan (Garland & Howard, 2009). So, what happens to your brain when you meditate? More importantly, how does scripture meditation help you grow in Christlikeness?

MINDFULNESS MEDITATION

Mindfulness-based meditation or mindfulness-based intervention has been the subject of increasing research interest in the field of psychology, neuroscience, health care, and education in recent years for its perceived benefits to various aspects of people's health. A large body of evidence has shown that mindfulness meditation produces a wide range of beneficial effects, including improvements in cognitive functioning (Lutz et al., 2009), psychological well-being (Luders et al., 2009), and physical health (Jacobs et al., 2011) for healthy participants as well as for those with clinical disorders such as anxiety, depression, memory loss (Lazar et al., 2000), and attention-deficit/hyperactivity disorder (Guastella et al., 2010). Moreover, mindfulness meditation has made its way into academia from elementary schools all the way to graduate schools (Paul, Elam, & Verhulst, 2007; Shapiro, Schwartz, & Bonner, 1998) where students receive instructions to practice mindfulness meditation to reduce anxiety, improve attention, and more. Though many attribute the concept of mindfulness to ancient Buddhist, Hindu, and Chinese philosophies, mindfulness can be characterized as a universal human capacity (Ie, Ngnoumen, & Langer, 2014; Kabat-Zinn, 2003, p. 146) that is also deeply rooted in Jewish, Islamic, and Christian religions (Trousselard, Steiler, Claverie, & Canini, 2014).

What is mindfulness? Today, a variety of definitions are offered. Kabat-Zinn (2013), defines mindfulness as *intentionally* paying *focused attention* to one's bodily sensations, feelings, thoughts, and/or any other internal and external experiences as they arise in the *present moment*, with *nonjudging*, *nonreactive, open,* and *accepting* attitude. Mindfulness, in general, simply means *"mindful awareness," being aware or mindful, without which, one is not able to live a purposeful life.* While researchers still debate about the exact operational definition of mindfulness itself, they generally agree that mindfulness involves intentionally bringing one's attention to the internal and external experiences occurring in the present moment and is often cultivated through a variety of meditation exercises (Baer, 2003), such as sitting meditation, walking meditation, or mindful movements (Kabat-Zinn, 2003, 2013).

One style of mindfulness meditation that is studied extensively by neuroscientists, and therefore will be discussed in this chapter, is *focused attention (FA) meditation*. In FA meditation one voluntarily focuses attention on a chosen object—such as a visual object, breathing sensation, or visualized image—for a sustained period of time and prevents one's attention from drifting from the chosen object (Lutz, Dunne, & Davidson, 2007). Since sustaining this focus also requires the practitioner to constantly monitor and regulate the quality of attention, FA meditation trains not only one's ability to sustain attention, but also develops other regulatory skills: (1) monitoring faculties that remains vigilant to distractions without destabilizing the intended focus; (2) disengaging from a distracting object; and (3) redirecting focus to the chosen object (Slagter, Davidson, & Lutz, 2011).

Defining *meditation* can also prove challenging since the term refers to a wide variety of practices and meditational practices vary from tradition to tradition. Neuroscientists Newberg and Iversen (2003) defined meditation as "a complex mental process involving changes in cognition, sensory perception, affect, hormones, and autonomic activity" (p. 282). As Baer (2003) notes, *mindfulness* is often taught and practiced through a variety of *meditation* exercises. In current neuroscience research context, *mindfulness, mindfulness practice, mindfulness training, mindfulness meditation,* and *mindfulness-based meditation* are being used interchangeably to refer to "*practicing mindfulness.*" Thus, in this chapter these terms will be used interchangeably. What appears to be a common characteristic is they all have "intentionally paying focused attention" on an object or a concept as a major component.

NEUROSCIENCE RESEARCH ON MINDFULNESS MEDITATION

Mindfulness neuroscience emerged in the 1990s as an increasing number of neuroscientists turned their attention to mindfulness meditation, particularly to Buddhist tradition (Davidson et al., 2003; Lu et al., 2014). Studies reporting on the practice of meditation from the Christian tradition are minimal. Newberg, who studied Franciscan nuns (2001) and Pentecostals (2006), as well as Beauregard, who studied Carmelite nuns (2007), represent two of the very few researchers who studied subjects engaging spiritual practices from the Christian tradition. Due to the lack of study done by neuroscientists on the effect of Christian meditation on the brain, I will review the studies done on the effect of mindfulness meditation, mostly from the Buddhist tradition. I will discuss how mindfulness impacts the brain, then I will discuss why I believe Christian meditation may also have a similar impact on the brain.

Different types of meditation activate or deactivate different parts of the brain. D'Aquili and Newberg (1999) demonstrated that meditation activates certain parts of the brain (e.g., the prefrontal cortex, which is critical for attention-focusing tasks and meta-awareness) while deactivating some other parts (e.g., areas in the parietal lobes that give a sense of orientation in space and time). Newberg and colleagues (2001) hypothesized that blocking all sensory and cognitive input into these areas during meditation is associated with the reduced sensation of space and time that is so often described in meditation. Newberg and Iversen (2003) then proposed a hypothesis outlining certain key parts of the brain that are likely to play important role in meditation. Since then, many more studies have documented that different types of meditation activate or deactivate different parts of the brain (Newberg et al., 2001; Newberg et al., 2003; Newberg & Waldman, 2009; Beauregard, 2007; Holzel et al., 2008; Holzel et al., 2011a; Luders et al, 2009; Luders et al., 2012a; Lutz et al., 2004; Petersen & Posner, 2012). In contrast to meditation, however, Newberg found no direct effect of *brief prayer* (e.g., few minutes) on cognition and speculated that it is the "intense, ongoing focus" that stimulates the cognitive circuits in the brain (Newberg & Waldman, 2009, p. 28).

Does Meditation Change Brain Structure?

While meditation types and measures differ between studies, results from the studies of long-term advanced meditation practitioners consistently demonstrated the association between meditation and *structural* changes of the brain regions that are typically activated during meditation (Holzel et al, 2008; Luders et al., 2009). For example, in Holzel and colleagues' study (2007), long-term advanced meditators, compared with nonmeditators, showed increased activation and greater cortical thickness in the anterior *cingulate cortex*, which is typically involved in executive attention—the ability to choose what to pay attention to and what to ignore. Also, greater gray matter (GM) volume and density were found in brain regions that are involved in introspection and in the regulation of

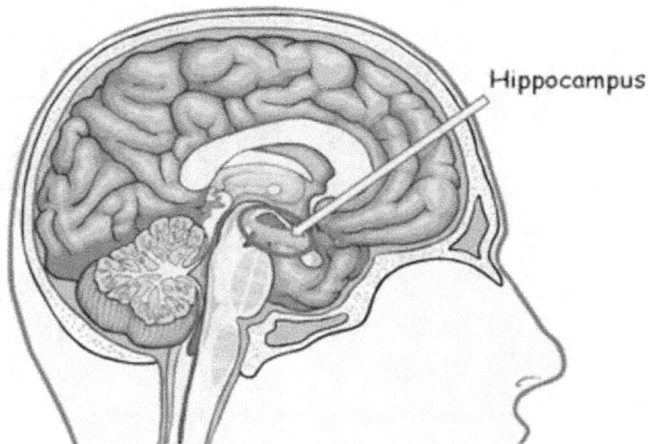

Figure 5.1.

arousal and emotions, particularly the *insula, hippocampus, prefrontal cortex, and brainstem*. GM refers to the areas throughout the brain that contain the cell bodies of a majority of the brain's neurons, and changes in GM reflect changes in brain morphology (structure) (see Figure 5.1).

Of particular importance is the increased GM in the hippocampus. The hippocampus is centrally located in the brain and has been known to play a key role in learning and memory processes (Squire, 1992) and has been postulated to play a central role in mediating some of the benefits of meditation (Newberg & Iversen, 2003). It is now shown to contribute to emotional regulation (Corcoran et al., 2005; Milad et al., 2007) (see Figure 5.2).

When the brain processes an emotional stimulus, it occurs through a direct path and an indirect path. The direct path (also known as "stream of feeling") involves the emotional stimulus triggering the thalamus which then communicates to the hypothalamus which then leads to a visceral, often instinctual, bodily response. In contrast, in the indirect path (also known as "stream of thinking"), the emotional stimulus is communicated via the thalamus to the sensory cortex and then to the cingulate cortex, where the stimulus is processed and an appropriate emotional response is formulated. This response is then communicated to the hippocampus for further processing and then communicated to the hypothalamus for execution of the body's response. This "stream of thinking" path thus allows for top down cortical control of emotional responses. Both the direct and indirect pathways are activated to various degrees in response to an emotional stimulus, and it is through the indirect pathway

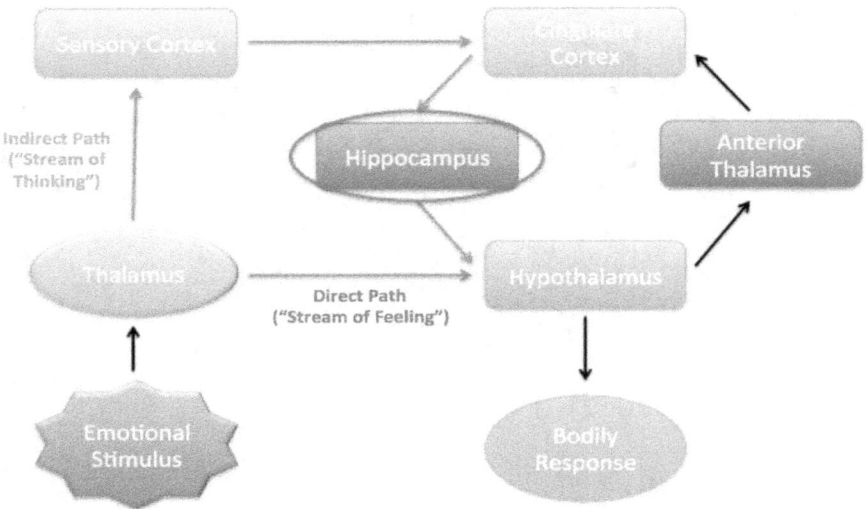

Figure 5.2.

that one is able to intentionally regulate one's emotions to a given stimulus and temper the body's instinctual response generated by the direct pathway. The increased GM in the cingulate cortex and the hippocampus of long-term meditators compared to nonmeditators seem to suggest that meditation is involved somehow in the development/growth of this indirect pathway.

Interestingly, decreased GM in the hippocampus has been found in several pathological conditions (e.g., major depression [Sheline, 2000] and posttraumatic stress disorder [Kasai et al., 2008]) including Alzheimer's diseases. In Alzheimer's diseases, the hippocampus is one of the first brain regions to suffer damage. Since the hippocampus is a region that is known for its ability to generate new neurons throughout one's lifespan, the volume loss in the hippocampus may be reversible in some conditions (Gage, 2002). It has been suggested that some of the behavioral effects of antidepressant treatment might depend on neurogenesis in the hippocampus (Santarelli et al., 2003). If that is so, then perhaps meditation that can induce neurogenesis in the hippocampus can be used in place of antidepressants. Moreover, the normal age-related decline in GM volume and in attentional performance was present in nonmeditating controls, but not in meditators (Ott, Holzel, & Vaitl, 2011, p. 119), suggesting meditation's role in slowing down the aging process. Also, Boyke and colleagues (2008) observed significant increase in GM volume in the hippocampi of the elderly when they were taught to learn a new skill like three-ball cascade juggling. Structural changes (e.g., GM changes) have been reported in as few as seven days, but whether that change is permanent is not clear (Ott, Holzel, & Vaitl, 2011, p. 125).

Long-Term Meditators Have Structural Changes Regardless of the Meditation Style

Luders and colleagues (2009) studied 22 active, long-term meditation practitioners. Subjects meditated using their own meditative style. Luders found a significantly larger volume of GM in the right hippocampus and orbitofrontal cortex of all meditation practitioners regardless of style. While the hippocampus, as previously discussed, contributes to emotional regulation, the orbitofrontal cortex, located in the frontal lobe of the brain, is involved in the cognitive processing of decision making. Therefore, greater volumes in these regions might account for "meditators' singular abilities and habits to cultivate positive emotions, retain emotional stability, and engage in mindful behavior." The authors suggest that "these regional alterations in brain structures constitute part of the underlying neurological correlate of long-term meditation *independent of a specific style and practice*" (Luders et al., 2009, p. 672, emphasis added). In other words, Luders and colleagues (2009) suggest that regardless of the specific meditation style employed, long-term meditators have *structural changes* in brain regions that are involved in *emotion regulation* and *response control*.

It has been proposed that the wide range of benefits associated with mindfulness meditation are produced by a number of different but interacting cognitive mechanisms—specifically, attention regulation, body awareness, emotion regulation, and change in self-perspective—that work synergistically to establish a process of enhanced self-regulation (Holzel et al., 2011a). Recent neuroimaging studies provide supporting evidence for these claims by demonstrating that mindfulness meditation practice improves self-regulation through modifying functions and *structures* of multiple parts of the brain that are involved in self-regulation (Holzel et al., 2011a).

While the changes to the hippocampus induced by meditation are emphasized in this chapter, it is important to know that several other areas of the brain are also found to exhibit significant changes. In a meta-analysis of 21 neuroimaging studies examining about 300 meditation practitioners, Fox and colleagues (2014) found seven other brain regions, hypothesized to be involved in meditation, consistently being altered in meditators. The regions include areas critical to meta-awareness (prefrontal cortex), exteroceptive and interoceptive body awareness (sensory cortices and insula), self and emotion regulation (anterior and mid cingulate; orbitofrontal cortex), and intra- and interhemispheric communication (superior longitudinal fasciculus; corpus callous). The most consistently altered brain regions in meditation practitioners, across all meta-analyses, include the following: prefrontal cortex, anterior/mid cingulate cortex, anterior insula, somatomotor cortices, inferior temporal gyrus, and hippocampus. Though causality cannot be assumed from the findings, Fox and colleagues conclude that meditation is consistently associated with changes in the brain structure.

Although mindfulness meditation neuroscience research is still in its infancy, it is evident that meditation practice is associated with structural changes in brain regions that are typically activated during meditation and in regions that are relevant to the practice of meditation. Even short-term (e.g., 8 week MBSR) training was found to induce structural changes (Holzel et al., 2011b), though it is unclear whether these changes reflect lasting changes in function and structure. Long-term meditators (e.g., 10,000 plus hours and 10 plus years), however, show very enduring "trait-like" transformation even at "baseline" or "default mode (Lutz, Slagter, & Davidson, 2008; Lutz, Dunne, & Davidson, 2007). It is suggested that while early stages of mindfulness meditation tend to reflect a more effortful, self-directed attentional and emotion-regulatory stance, long-term meditators meditate with more efficiency and less effort (Lutz, Slagter, & Davidson, 2008).

MECHANISMS BY WHICH MINDFULNESS MEDITATION ALTERS THE BRAIN STRUCTURE

How does the practice of mindfulness meditation alter the brain structure? Neuroplasticity and focused attention (FA) suggest the answer. Through the plasticity of the brain, can practicing mindfulness meditation alter the structure of the brain? In the previous section, we learned that neuroplasticity defines the ability of the brain to change its structure in response to experience. FA, which is a major component in mindfulness meditation, entails a form of experience we can use to change the physical structure of the brain.

Altering Brain Structure: Power of Focused Attention (FA)

What is the mind? Though the mind can be defined in various ways, Siegel's definition, albeit from a purely naturalistic view, proves helpful in understanding the process of altering the structure of the brain. The mind is "an embodied and relational process, emerging from within and between brains, that *regulates* the *flow of energy* and *information*" (Siegel, 2012a, p. 3). Siegel's definition of the mind helps us to understand why we so often find our mind wandering and how we can overcome it. *"[F]low* of energy and information" suggests that unless we put effort to focus, our mind will continually flow and wander. Attention is "the *process* that shapes the direction of the *flow* of energy and information" (Siegel, 2012b, pp. 7–11) and *focused attention* refers to a mental state in which the mind is unwaveringly and clearly focused (Lutz, Dunne, & Davidson, 2007).

Neurons That Fire Together, Wire Together

Siegel likens "attention" to the "scalpel" that a surgeon uses to perform surgery. He writes, "Attention is to a clinician or teacher what a scalpel is to a surgeon" (2012b, pp. 7–14). By *intentionally focusing our attention on an object/idea*, we will activate neurons in certain part of the brain that are responsible for processing that information. With repeated activation of those neurons over time, the connection between those neurons grows stronger. Similarly, neuronal pathways that are not utilized become weaker and weaker over time. This process is known as "Hebbian Law" named after Donald Hebb, and it means that the neural firing patterns will change because of *focused attention* (Siegel, 2012b). The mechanism by which neuronal connection is strengthened is as follows:

In any given neuronal pathway in the brain, there are many neurons that are in communication with each other. The way a neuron communicates to its neighboring neuron is through a chemical molecule called a neurotransmitter. The strength of the connection between one neuron to the next neuron depends on two factors: how many neurotransmitters the neuron sends as a signal to the next neuron and how many receptors the neuron receiving the signal has to receive the neurotransmitters. The more neurotransmitters released and the more receptors present, the stronger the connection and the stronger a signal is relayed (see Figure 5.3).

Figure 5.3 shows that when a neuron is in frequent communication with another neuron (i.e., through repeated stimulation), a physiologic process occurs called "long-term potentiation," whereby the capacity of the sender neuron (top neuron in the diagram) to send neurotransmitters is increased as well as the number of neurotransmitter receptors on the receiving neuron (bottom neuron). Thus, repeated stimulation of those neurons strengthens the connection as each subsequent stimulus elicits a stronger signal between the neurons. Moreover, the increase in

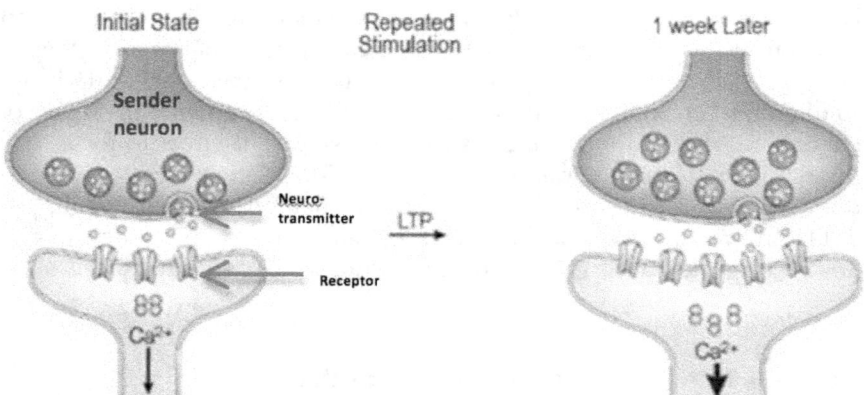

Link: https://sites.google.com/site/mcauliffeneur493/home/synaptic-plasticity

Figure 5.3.

receptors, neurotransmitters, and proteins devoted to the maintenance of highly utilized neurons eventually contribute to the increased GM of cortical areas previously mentioned to be seen in long-term meditation practitioners.

Neurons that are activated together (i.e., fire together) once are likely to be activated together again in the future. If we *focus attention* repeatedly, neurons that are activated together repeatedly will develop connections and rewire, leading to the phrase *"neurons that fire together, wire together."* On a more global scale, the strengthening of different neuron connections in our brain through repeated use while weakening of other neuron connections through neglect leads to the reshaping of the brain. In other words, what we do with our *mind* can change the very structure of our brain. Moreover, the more *intense* and *specific* our neural firing (which we create with the *focus of our attention*), the more specific we can be in directing neural pathways in the direction we want and the more long-lasting synaptic changes we are likely to initiate in our brain (Siegel, 2012b). In short, it is through the process of *focusing attention* and *repetition* (which leads to *neurons that fire together, wire together*) that mindfulness is associated with change in the brain structure. The more *intense* and *repeated* the mindfulness practice, the more lasting the change will be (Slagter et al., 2011; Newberg & Waldman, 2009).

What does all this mean for spiritual formation? Can scripture meditation and reflection also induce change in brain structure as mindfulness meditation can? The next section explores these questions.

SCRIPTURE MEDITATION AND REFLECTION

Throughout the Scripture, God's children are instructed to meditate (Josh. 1:8; Ps. 1:2; cf. Phil. 4:8–9). In the Old Testament, *haghah* and *siach* are two Hebrew words most often used to convey the concept of meditation. *Haghah* means "growl," "utter," or "moan," as well as "reflect," "ponder," or "muse" (Negotia, 1974, p. 322; Hartley, 1986, p. 305; Saucy, 2013, p. 143) and *siach* means to "ponder deeply" or "contemplate" (Gen. 24:63; Ps. 104:34). Meditation, therefore, involves *deep thinking, reflection,* and *contemplation*. However, in spite of God's instruction to meditate, meditation has become a neglected spiritual discipline by evangelical Christians over the last two centuries. As Packer (1973) observes, "meditation is a lost art today and Christian people suffer grievously from their ignorance of the practice" (p. 23). Saucy (2013) describes meditation as "engaging in deep thought to the extent that it touches one's *whole being*" (p. 153). As noted, neuroscientists Newberg and Iversen's view of meditation as "a complex mental process involving changes in cognition, sensory perception, affect, hormones, and autonomic activity" (2003, p. 282) also indicates meditation's involvement of and effect on a person's entire being.

A Profoundly Integrative, Holistic, Dynamic, and Relational Process

Scripture meditation, however, cannot be defined fully without also defining Christian prayer, because they are inseparably interrelated. Broadly defined, prayer is a form of bidirectional communication wherein not only a person who prays expresses and communicates one's thoughts, emotions, requests, and other such ideas to God, but God also speaks to the person through various means (e.g., thoughts, images, sound, etc.). People often understand prayer as petitioning, the unidirectional presenting of requests to God, yet this practice remains but a small aspect of prayer. Christian prayer defines a *dynamic* and *relational* process in which both parties, the person who prays and God, speak and listen in the context of a loving relationship initiated by God.

Scripture meditation can be described as a form of prayer or bidirectional communication, wherein we intentionally focus our attention on God's Word and invite God to speak to us through scripture. As God speaks, we receive and hear God's message, meditate on God's message, reflect on how it touches our lives, respond to and communicate with God in the form of obedience to God's message, and submit our petitions in prayer. This type of scripture meditation is known as *lectio divina* (Latin for divine reading or sacred reading), which has been practiced by early Christians from the third century.

From a meditator's perspective, scripture meditation, if done properly, defines a profoundly *integrative* and a *holistic* process. A process that involves and *integrates* one's *entire being* entailing thinking, feeling, and will (i.e., desires and choices) as one encounters God personally through God's Word. Compared to mindfulness meditation, scripture meditation is profoundly holistic in many aspects. For instance, with respect to time, scripture meditation is not limited to the present moment of meditator's life, but may encompass the past, present, and the future, although focus may be on the present moment. With respect to the meditator's life, for example, scripture meditation is not limited to any particular component of the meditator's life but may encompass all components of meditator's life as the Holy Spirit guides the process and enlightens the mind of the meditator to pay attention to various areas in life (for example, in areas of finance management, eating and drinking habits, relationship with others, etc.). In short, through scripture meditation, the meditator's entire being (not only cognitive, but also affective and volitional aspect), entire life (not only the present but also the past and future), and all components of the meditator's life can be addressed and thus affected. Scripture meditation, therefore, is a profoundly *integrative, holistic, dynamic,* and *relational* process wherein both the meditator and the object of the meditation, God, are engaged in interaction, giving and receiving messages.

Focused Attention (FA) Continually: Fix Our Eyes on Jesus

In scripture meditation, the object of meditation is God, and we are to *"fix"* our eyes *(focus our attention)* on *Jesus* (cf. Heb. 12:2), though we also reflect on our "self" and our lives in light of our growing knowledge of and *relationship* with Him. Although the word "meditate" is generally not found in many modern translations of the New Testament, the concept of meditation can be seen in Paul's letter to the Philippians. Paul exhorts to the Philippians to "dwell on" (4:8, NASB) or "think about" (NIV, ESV) things that are true, noble, right, pure, lovely, admirable, and so on. From the Greek text, O'Brien (1991) translates that verse as "let your thoughts *continually dwell on* these things [so that your conduct will be shaped by them]" (p. 415). Both the Old Testament and New Testament instruct believers to focus attention on God, God's deeds, and things of God (e.g., Josh. 1:8; Heb. 12:2; cf. Phil. 4:8) and *meditate* on God's Word not only repeatedly but *continually* so that they will live according to God's Word (Josh. 1:8-9; Ps. 1: 1; cf. 1Thess. 5:17) and not according to what they perceive from the world (2 Cor. 5:7). In short, intentionally paying *focused attention continually* is a major component in scripture meditation as *focused attention* and *repetition* are the major components in the mindfulness meditation alongside other key elements.

Reflection, Attention, Christlikeness, and Training

Reflection remains an integral part of Scripture meditation, as the Hebrew word *hagah* implies. We are to reflect on whatever the Spirit of God will bring to our mind. In mindfulness meditation, the meditator is the one who is constantly taking charge and directing the process to focus attention on oneself or a given object/concept in the present moment. However, in Scripture meditation, although we may intentionally initiate the meditation by focusing on certain Scripture text, what may actually happen during the meditation is left open. The Holy Spirit may guide us in the direction He wants as we remain sensitive to His guidance, as we *focus our attention* on Him who is not an object or concept but a relational person. Thus, scripture meditation and reflection is a *dynamic* process wherein both the meditator and the Holy Spirit interact, speak (not necessarily verbally), and listen.

In mindfulness meditation, a meditator's constant goal is to *focus attention* on *one's self* and anything that is arising at the *present moment*. In Scripture meditation, *reflection* should encompass not only the present, but also the *past and future*. Though we are instructed to live in the moment, we are also instructed to *"remember"* especially the deeds of God from the past, including in our lives. Moreover, we are instructed to live each moment with a great anticipation of Christ's coming in the future. Siegel (2012b) suggests that it is the *hippocampus* that acts as a "search engine" of memory retrieval as well as a "master puzzle assembler" by integrating the scattered pieces of unreflected life's events into a coherent whole story through reflection process, thereby increasing self-understanding which, in turn, contributes to growth.

Our *goal* in Scripture meditation and Christian life is to grow in *deeper knowledge of Him and deepen our relationship with Him* so that we will be conformed to increasing *likeness of Jesus Christ* (Rom. 8:29; cf. Eph. 4:13; Gal. 4:19). Reduction of anxiety, stress, worry, and enhanced ability to love, self-regulate, and so on (i.e., the benefits neuroscience researchers associated with mindfulness meditation practice, predominantly from the Buddhist tradition) reflects the characteristics of our life (fruit of the Holy Spirit) that will be developed and produced *as we meditate and grow in our relationship with Christ*. Those benefits themselves are not the ultimate goal, but are the manifestation of deepening of our personal relationship with Christ. Scripture meditation may include the concepts of sufferings and trials since the process of Christian growth assumes *suffering and trials* (cf. John 16:33; Job 23:10; Phil. 2:5–8; Matt. 16:24–25). We are to endure because Christ Himself became a model for us by enduring suffering and trials (Heb. 12:2–3, 11). We will grow in Christ-like character through the processes, which most likely will include trials and suffering (Jas. 1:2–4; Ps. 119: 67, 71; Acts 14:22).

We are instructed to *meditate* on God's Word not only repeatedly but *continually* so that we will live according to God's Word (Josh. 1:8–9; Ps. 1:1; cf. 1Thess. 5:17) and not according to what we perceive from the world (2 Cor. 5:7). Engaging in Scripture meditation includes a *training* that needs to be developed, and it calls for *desire* and *intentionality*, without which it will not happen (cf. 1Tim 4:7–8; Willard, 2002).

CHANGING BRAIN STRUCTURE THROUGH SCRIPTURE MEDITATION AND REFLECTION

The practice of mindfulness meditation induces structural changes in the brain through the mechanism of *focused attention (FA)* and *repetition*. Can the practice of Scripture meditation and reflection also change the brain structure? *Focused attention* and *repetition* are also essential characteristics of Scripture meditation and reflection. Kabat-Zinn (2003), quoting Thera (1962), claims that "historically, mindfulness has been called 'the heart' of Buddhist meditation." I would equally argue that *mindfulness* has been "the heart" of Christian meditation as well, even though they are different types of meditation. Christians are instructed to be mindful of God continually and to live in awareness of that knowledge of His presence.

Scripture meditation and reflection not only possesses all the essential attributes of mindfulness meditation found to be associated with altering the structure of the brain (i.e., focused attention, repetition, intention), but it has even more positive attributes (e.g., personal encounter with a *loving* God who gives *purpose* for life, *hope* for the future, and challenges Christians to live by *faith and not by sight*). According to neuroscience, having loving relationships, hope, faith, and purpose for life are all very good for brain health (Siegel, 2012b; Newberg & Waldman, 2009).

CHANGING CORE BELIEFS THROUGH SCRIPTURE MEDITATION AND DEEP REFLECTION

We have heard that we live according to what we truly believe. We live according to what we truly believe, not what we profess to believe. According to Christian philosopher Moreland (1997), "Beliefs are the rail upon which our lives run" and "almost always we act according to what we really believe," that is, our *core beliefs*, not professed beliefs (p. 73). He further states that, "A belief's impact on behavior is a function of three of the belief's traits: its content, strength, and centrality" (p. 73). The content of belief refers to what we believe and the strength of belief refers to the degree to which we are convinced the belief is true. As we gain more "*evidence* and *support* for a belief, its strength grows" for us that it may move from, for example, "fairly likely, quite likely, beyond reasonable doubt" to "completely certain." "The more certain [we] are of a belief, the more it becomes a part of [our] very soul, and the more [we] rely on it as a basis for action" (p. 73). The last trait, centrality of belief, refers to the degree of the belief's importance in relation to all other beliefs we hold. For example, my belief that kale is good for my health is very strong, but that belief is not very central in my overall belief system and consequently, I may not act on that belief (not eat kale) even when I am strongly convinced that kale is good. On the contrary, if the belief I hold is very central in my overall belief system, I am more likely to act on it. Core beliefs are those beliefs that are held strongly and are central in our overall belief system.

Spiritual formation into Christlikeness involves changing our core beliefs. So how do we change our core beliefs so that its content reflects what God wants us to believe and its strength grows to "completely certain" and the belief becomes very central in our belief system that we will act on that belief? Moreland asserts that we cannot change our core beliefs directly at our will, but it is possible to change them indirectly through, for example, engaging in activities of the *mind* such as *study, meditation*, and *reflection* that would change the *content, strength*, and *centrality* of our belief which, in turn, will "transform our character and behavior" (Moreland, 1997, p. 75). Scripture also seems to teach the transformation of *the mind* as a first step for the spiritual formation of a whole person (cf. Ro 12:2). But how does engaging in meditation, which is an activity of the mind, change our *core beliefs*?

Thoughts Meditated Deeply and Repeatedly for a Long Time Become Inner Reality

Newberg and colleagues found, in nearly all of their subjects who had meditated for over 10 years, asymmetric activity between the left and right half of the thalamus when they were not engaged in any meditation (Newberg et al., 2001; Newberg & Iversen, 2003). One side was more active than the other side when in the general population, both sides are typically equal in activity, especially when they are at rest (Luders et al.,

2009 also found asymmetrical thalamic activity in long-term active meditators with more than 20 years of meditation). The thalamus is believed to play an important role of identifying what is real and what isn't. Newberg and Waldman argue that "the more you meditate on a specific object—be it God, or peace, or financial success—the more active your thalamus becomes, until it reaches a point of stimulation where it perceives thoughts in the same way that other sensations are perceived" (2009, p. 55).

If you meditate repeatedly for a long period of time, your brain will begin to respond as though the idea you meditated on is reality. In other words, thoughts, *if meditated upon repeatedly for a long period of time, become an inner reality and the "normal state of awareness"* for the meditator as real as what they perceive through senses. For advanced meditators, M. Newberg and colleagues (2009) argue that this (meditated thoughts becoming a reality and a normal state of awareness), is what is taking place. "Thus, the more you focus on God, the more God will be sensed as real" (p. 55) and one's inner reality that has been altered through meditation can be different from what one perceives through one's senses, outer reality. Whether the thalamus is the major part of the brain that is responsible for changing inner reality or not is something that needs to be studied further. What can be gleaned from the findings, however, is that the more we meditate on the Scripture (or anything), the more real it becomes to us. This seems to support Moreland's claim that our beliefs (the content, strength, and centrality) can be changed through *intentional, repetitious Scripture meditation* and *reflection*.

Newberg and colleagues' findings about *deeply meditated ideas becoming an inner reality (core beliefs)* for a meditator confirm my personal experience as a long-term meditator. Over the years, I have experienced that what I have meditated on have become my deep beliefs (core beliefs) and inner reality, and actually have become stronger than what I perceive through my senses from the world. Consequently, as Moreland said, it has become more effortless and more natural to live out from what I meditate on even when it is different from what I perceive through my senses.

Newberg and colleagues' findings also seem to confirm God's design of the human brain to accomplish His purpose, that is, conforming people into Christlikeness (Rom. 8:29— For those God foreknew he also predestined to be conformed to the likeness of his Son). Scripture teaches us to be transformed by the renewing of our *mind* (Ro 12:2), and mindfulness neuroscience research findings show it is through *focused attention* (focusing mind) and *repetition* that we change our brain structure. God's instruction to His children to *focus their attention* on Jesus (Heb. 12:2), *meditate* on Him and His Word day and night (Josh. 1:8), and *think about*, for example, "[w]hatever is true, whatever is noble, whatever is right, whatever is pure, whatever is lovely, whatever is admirable—if anything is excellent or praiseworthy" (Phil. 4:8) all seem to help us understand why God might have instructed us to do above things, for they will reshape our brain structure and change our inner reality (core beliefs) in ways that will resemble the inner reality of Christ and facilitate our spiritual formation into increasing Christlikeness.

Hebrews 11:1 and 2 Corinthians 5:7

When I was reading the findings of Newberg and colleagues, Hebrews 11:1 and 2 Corinthians 5:7 came to my mind: "[F]aith is being sure of what we hope for and certain of what we do not see" and "We live by faith, not by sight." According to the findings from mindfulness neuroscience research on long-term meditators, Scripture meditation and reflection, if we practice continually long term, is likely to change our inner reality (core beliefs) according to the teachings of the Scripture because our brain will take what we meditated on as real, even if it differs from what we perceive from the outer world. When our inner reality developed by Scripture meditation is stronger than what we perceive through our senses from the world (i.e., when the strength of our core beliefs is increased), it would become that much easier for us to live according to the Scripture. We would then be living out Hebrews 11:1 and 2 Corinthians 5:7 because we would be certain of, for example, God's promises, even if we do not see. We will live by what we meditated on (which has become strong core belief) and not by what we see in the world. Scripture meditation is powerful in bringing about the transformation of core beliefs in the depth of one's heart.

NEUROPLASTICITY AND SPIRITUAL FORMATION INTO CHRISTLIKENESS

What does neuroplasticity have to do with spiritual formation? It is because the brain is plastic and malleable that we can change the very structure of our brain and also core beliefs through experience and training, such as meditation and reflection. The question, therefore, is no longer is it possible to grow in increasing Christlikeness, but rather how much do you really *desire* to grow in Christlikeness? Because if you so desire, you can be intentional in choosing what to think about with your mind. According to the words of neuroscientists Merzenich and deCharm, what we choose to think about and meditate on each moment will determine who we become in the future: "moment by moment we choose and sculpt how our ever-changing minds will work, we choose who we will be the next moment in a very real sense, and these choices are left embossed in physical form on our material selves" (1996, p. 76). I believe that God, who designed the human brain to work in such an amazing way, has left that choice of what to think about with one's mind in the heart of each individual.

IMPLICATIONS OF MINDFULNESS NEUROPLASTICITY RESEARCH FOR CHRISTIAN EDUCATORS

Though mindfulness neuroscience is still in its infancy, the initial findings that meditation can induce the rewiring of our neural network and change the structure of the brain have key implications for Christian educators. By training the mind of students, we can help them reshape their brain (i.e., change neural firing) in the direction of spiritual formation into increasing Christlikeness.

Be mindful. Teaching students to be *mindful*, that is, to be *mindfully aware* of their thinking and feeling as they live each moment, for example. Teach them to ask a simple question such as, *what am I thinking now? Is it good? If not, then change it.* Helping students to become aware of their own thinking can help them to recognize their wandering mind and choose to refocus, if they so desire. Also, it could help them to recognize unhealthy thought patterns and change. They could also ask questions regarding their feelings, such as, *what am I feeling now? Is it good? If not, then change it.* However, as Willard notes (2002), we cannot change our feelings directly at our will. But the good news is that we can change our feelings by, for example, changing our thoughts. So, if we teach students to pay attention to their feelings and how they can change negative feelings by directing their attention to think about something good, positive feelings will follow. For instance, when students become aware that they are feeling anxious, they can intentionally direct their attention to good thoughts, such as Philippians 4:6, which says, "Do not be anxious about anything, but in everything" and subsequently experience the anxiety subside. By teaching students to be *mindfully aware* of their thinking and feeling on an ongoing basis, we will be training their minds to focus and helping them to make *conscious choices* rather than living their life on an autopilot wherein their thoughts and feelings are dictated by their past (i.e., neural firing patterns of the past). Thus, living mindfully is the key to spiritual formation, for when individuals are mindfully aware of their thinking and feeling, they can exercise their choice and if they so desire, can choose what is good for them.

Teach how to meditate and practice meditation corporately and individually. Teachers can be intentional in teaching students how to meditate. Teaching students to focus their attention would be an invaluable tool. I find memorizing Scripture verses and passages to be very powerful and life-giving in many respects, especially to focus my attention on God when my mind wanders and to be aware of His presence throughout the day. Teachers can carefully select Scripture verses and passages according to the needs and abilities of their students and help them understand the meaning and put them into memory so that they can access the Scripture at any time they choose to focus their attention on God and His Word. This can be practiced corporately in class and individually outside of class. Also, centering prayers and short prayers can be effective in assisting individuals to focus their attention on God.

When students experience the benefits of meditation, they are more likely to practice on their own. I would suggest that teachers make meditation a part of their class, assignment, and curriculum and help students to practice repeatedly long term. For example, teachers can take a few minutes at the beginning of each class period and practice short meditation with the students that will help them be mindful and focus their attention. On a practical level, teachers can also give meditation assignments and develop curriculum with the goal of moving students from novice meditators to advanced meditators.

Take Time to Reflect. Teach and train students to take time to reflect on their life. Reflection helps the brain to integrate the isolated pieces of life's events into a coherent whole story that is understandable by the self which, in turn, increases self-understanding and contributes to growth.

Teach older adults novel things and skills. Generally speaking, the educational ministry in church is geared toward children, youths, and young and mid adults. Older adults are more often than not left out of active learning environments, and ministries to older adults are often limited to recreation, social activity, and entertainment. However, neuroplasticity occurs at all ages, including older adults. Research shows that when given a new stimulus, experience, or challenge, even older adults' brains continue to develop new neural connections. Teachers and ministers of older adults should not assume that older adults would not or could not learn new things, but instead should inspire them, encourage them to learn novel things, and teach them focused attention skills, meditation, and give new challenges that will stimulate neural growth in their brain and improve overall health.

Inspire by living mindfully and meditatively. Lastly, students will be more inspired and develop a stronger desire to learn and practice mindfulness (paying focused attention) and Scripture meditation and reflection if they can see the benefits of that being lived out in their teacher's life or others in faith community. Therefore, the application for the teachers could be that they become more mindful and aware of their own thinking and feeling and practice mindful living. Furthermore, engaging in Scripture meditation and reflection regularly in the long run would be an excellent way to grow in increasing Christlikeness and inspire students to do the same.

CONCLUSION

In this chapter I argued that *mindfulness* is not a concept that is only characteristic of Buddhist meditation, but is a universal human capacity which is also at the heart of Christian meditation. I also proposed that Scripture meditation and reflection not only possess all the essential attributes of mindfulness meditation found to be associated with structural changes of the brain and various health-related benefits, but also are a profoundly integrative, holistic, dynamic, and relational process wherein a meditator's entire being—both material body and immaterial mind/spirit/soul—is engaged, encountering God in a relationship initiated by a loving God. Finally, based on the initial findings from mindfulness neuroscience research and current understandings of the brain, I suggested that through intentional training, such as intensive, repetitious Scripture meditation, and deep reflection, we can alter not only the structure of the brain but also alter our inner reality (core beliefs) in such a way that our brain and inner reality will help us achieve our goal of spiritual formation into Christlikeness.

DISCUSSION QUESTIONS

1. What is neuroplasticity and how can our understanding of the neuroplasticity help us in the spiritual formation process?
2. What are the direct and indirect pathways through which the brain processes an emotional stimulus? How does your understanding of the pathways affect your view of meditation and prayer?
3. Compare and contrast Scripture meditation with mindfulness meditation. Why call scripture meditation holistic?
4. How does paying focused attention change brain structure? Explain the mechanism.

5. Moreland identifies "content, strength, and centrality" as the three traits of a belief. How can Scripture meditation and reflection change our core beliefs in such a way that they will facilitate the process of spiritual formation into increasing Christlikeness?
6. Regarding spiritual formation into Christlikeness, the author of this chapter states, "The question, therefore, is no longer is it possible to grow in increasing Christlikeness, but rather how much do you really *desire* to grow in Christlikeness?" What does the author mean by this? What are your thoughts on the statement? How do you create that *desire*?

REFERENCES

Baer, R. (2003). Mindfulness training as a clinical intervention: A conceptual and empirical review. *Clinical Psychology: Science and Practice, 10*(2), 125–143.

Begley, S. (2006). *Train your mind, change your brain: How a new science reveals our extraordinary potential to transform ourselves*. New York, NY: Ballantine.

Beauregard, M. (2007). *The spiritual brain: A neuroscientist's case for the existence of the soul*. New York, NY: Harper One.

Boyke, J., Driemeyer, J., Gaser, C., Buchel, C., & May, A. (2008). Training-induced brain structure changes in the elderly. *The Journal of Neuroscience, 28*(28), 7301–7305.

Corcoran K. A., Desmond T. J., Frey, K. A., & Maren, S. (2005). Hippocampal inactivation disrupts the acquisition and contextual encoding of fear extinction. *Journal of Neuroscience, 25*(39), 8978–8987.

dA'quili, E. G., & Newberg, A. (1999). *The mystical mind: Probing the biology of religious experience*. Minneapolis, MN: Fortress Press.

Davidson, R., Kabat-Zinn, J., Schumacher, J., Rosenkranz, M., Muller, D., Santorell, S., ... Sheridan, J. (2003). Alterations in brain and immune function produced by mindfulness meditation. *Psychosomatic Medicine, 65*, 564–570.

Eriksson, P., Perfilieva, E., Bjork-Eriksson, T., Alborn, A., Nordborg, C., Peterson D. & Gage, F. (1998). Neurogenesis in the adult human hippocampus. *Nature Medicine, 4*(11), 1313–1317.

Fox, K., Nijeboer, S., Dixon, M., Floman, J., Ellamil, M., Rumak, S., Sedlmeier, P., & Christoff, K. (2014). Is meditation associated with altered brain structure? A systematic review and meta-analysis of morphometric neuroimaging in meditation practitioners. *Neuroscience and Biobehavioral Reviews, 43*, 48–73.

Gage, F. H. (2002). Neurogenesis in the adult brain. *Journal of Neuroscience, 22*(3), 612–613.

Garland, E. & Howard, M. (2009). Neuroplasticity, psychosocial genomics, and the biopsychosocial paradigm in the 21st century. *Health & Social Work, 34*(3), 191–199.

Guastella, A., Einfeld, S., Gray, K., Rinehart, N., Tonge, B., Lambert, T., & Hickie, I. (2010). Development of inhibitory control in attention-deficit/hyperactivity disorder. *Biological Psychiatry, 67*(7), 692–694.

Hartley, J. E. (1986). Meditate. In G. W. Bromiley (Ed.), *The international standard Bible encyclopedia* (Vol. 3: K–P, pp. 305–306). Grand Rapids, MI: Eerdmans.

Holzel, B., Carmody, J., Vangel, M., Congleton, C., Yerramsetti, S., Gard, T., & Lazar, S. (2011a). Mindfulness practice leads to increases in regional brain gray matter density. *Psychiatry Research: Neuroimaging, 191,* 36–43.

Holzel, B., Lazar, S., Gard, T., Schuman-Oliver, Z., Vato, D., & Ott, U. (2011b). How does mindfulness meditation work? Proposing mechanisms of action from a conceptual and neural perspective. *Perspectives on Psychological Science 6*(6), 537–559.

Holzel, B., Ott, U., Gard, T., Hempel, H., Weygandt, M., Morgen, K., & Vaitl, D. (2008). Investigation of mindfulness meditation practitioners with voxel-based morphometry. *Social Cognitive and Affective Neuroscience, 3,* 55–61.

Holzel, B., Ott, U., Hempel, H., Hackl, A., Wolf, K., Stark, R., & Vaitl, D. (2007). Differential engagement of anterior cingulate and adjacent medial frontal cortex in adept meditators and non-meditators. *Neuroscience Letters, 421,* 16–21.

Jacobs, T., Epel, E., Lin, J., Blackburn, E., Wolkowitz, O., Bridwell, D., Zanesco, A., & Aichele, S. (2011). Intensive meditation training, immune cell telomerase activity, and psychological mediators. *Psychoeuroendocrinology (36),* 664-681.

Ie, A., Ngnoumen, C., & Langer, E. (Eds.). (2014). *The Wiley Blackwell handbook of mindfulness* (Vol. 2). New York, NY: Wiley & Sons.

Kabat-Zinn, J. (2003). Mindfulness-based interventions in context: Past, present, and future. *Clinical Psychology: Science and Practice, 10*(2), 144–156.

Kabat-Zinn, J. (2013). *Full catastrophe living: Using the wisdom of your body and mind to face stress, pain, and illness.* New York, NY: Bantam.

Kasai, K., Yamasue, H., Gilbertson, M. W., Shenton, M. E., Rauch, S. L., & Pitman, R. K. (2008). Evidence for acquired pregenual anterior cingulated gray matter loss from a twin study of combat-related post-traumatic stress disorder. *Biological Psychiatry, 63*(6), 550–556.

Lazar, S., Bush, G., Gollub, R., Fricchione, G., Khalsa, G., & Benson, H. (2000). Functional brain mapping of the relaxation response and meditation. *Neuroreport, 11*(7), 1581–1585.

Lu, H., Song, Y., Xu, M., Wang, X., Li, X., & Liw, J. (2014). The brain structure correlates of individual differences in trait mindfulness: A voxel-based morphometry study. *Neuroscience, 272,* 21–28.

Luders, E., Kurth, F., Mayer, E., Toga, A., Narr, K., & Gaser, C. (2012). The unique brain anatomy of meditation practitioners: Alterations in cortical gyrification. *Frontiers in Human Neuroscience, 6*(34). Retrieved from http://www.ncbi.nlm.nih.gov/pmc/articles/PMC3289949/

Luders, E., Phillips, O., Clark, K., Kurth, F., Toga, A., & Narr, K. (2012). Bridging the hemispheres in meditation: thicker callosal regions and enhanced fractional anisotropy (FA) in long-term practitioners. *Neuroimage, 61*(1), 181–187.

Luders, E., Toga, A., Lepore, N., & Gaser, C. (2009). The underlying anatomical correlates of long-term meditation: Larger hippocampal and frontal volumes of gray matter. *Neuroimage, 45*(3), 672–678.

Lutz, A., Dunne, J., & Davidson, R. (2007). Meditation and the neuroscience of consciousness: An introduction. In P. Zelazo, M. Moscovitch, & E. Thompson (Eds.), *Cambridge handbook of consciousness* (pp. 81–109). New York, NY: Cambridge University Press.

Lutz, A., Greischar, L., Rawlings, N., Ricard, M., & Davidson, R. (2004). Long-term meditators self-induce high-amplitude gamma synchrony during mental practice. *Proceedings of the National Academy of Sciences of the United States of America, 101*(46), 16369–16373.

Lutz, A., Slagter, H., & Davidson, R. (2008). Cognitive-emotional interactions: Attention gulation and monitoring in meditation. *Trends in Cognitive Science, 12*(4), 163–169.

Lutz, A., Slagter, H., Rawlings, N., Francis, A., Greischar, L, & Davidson, R. (2009). Mental training enhances attentional stability: Neural and behavioral evidence. *The Journal of Neuroscience, 29*(42), 13418–13427.

Merzenich, M. M., & deCharms, R. C. (1996). Neural representations, experience, and change. In R. R. Llinas & P. S. Churchland (Eds.), *The mind-brain continuum: Sensory processes* (pp. 61–82). Cambridge, MA: MIT Press.

Milad, M. R., Wright, C. I., Orr, S. P., Petman, R. K., Quirk, G. J., & Rauch, S. L. (2007). Recall of fear extinction in humans activates the ventromedial prefrontal cortex and hippocampus in concert. *Biological Psychiatry, 62*(5), 446–454.

Moreland, J. (1997). *Love your God with all your mind: The role of reason in the life of the soul*. Colorado Springs, CO: Navpress.

Negotia, A. (1974). Hagah. In G. J. Botterweck, H. Ringgren, & Heinz-Josef (Eds.), D. E. Green (Trans.), *Theological dictionary of the Old Testament* (Vol. III, pp. 321–324). Grand Rapids, MI: Eerdmans.

Newberg, A., Alavi, A., Baime, M., Pourdehnad, M., Santanna, J., & d'Aquili, E. (2001). The measurement of regional cerebral blood flow during the complex cognitive task of meditation: A preliminary SPECT study. *Psychiatry Research: Neuroimaging, 106*, 113–122.

Newberg, A. & Iversen, J. (2003). The neural basis of the complex mental task of meditation: Neurotransmitter and neurochemical considerations. *Medical Hypothesis, 61*(2), 282–291.

Newberg, A., Pourdenhnad, M., Alavi, A., & Aquili, E. (2003). Cerebral blood flow during meditative prayer: Preliminary findings and methodological issues. *Perceptual and Motor Skills, 97*, 625–630.

Newberg, A., & Waldman, M. (2009). *How God changes your brain*. New York, NY: Ballantine Books.

Newberg, A., Wintering, N., Morgan, D., & Waldman, M. (2006). The measurement of regional cerebral blood flow during glossolalia: A preliminary SPECT study. *Psychiatry Research: Neuroimaging, 148*, 67–71.

O'Brien, P. (1991). *The epistle to the Philippians: A commentary on the Greek text*. Grand Rapids, MI: Eerdmans.

Ott, U., Holzel, B., & Vaitl, D. (2011). Brain structure and meditation: How spiritual practice shapes the brain. *Studies in Neuroscience, Consciousness and Spirituality, Springer,* (1), 119–128.

Packer, J. I. (1973). *Knowing God*. Downers Grove, IL: Inter Varsity Press.

Paul, G., Elam, B., & Verhulst, S. (2007, Summer). A longitudinal study of students' perceptions of using deep breathing meditation to reduce testing stresses. *TEACH LEARN MED, 19*(3), 287–292.

Petersen, S., & Posner, M. (2012). The attention system of the human brain: 20 years after. *Annual Review of Neuroscience, 35,* 73–89.

Santarelli, L., Saxe, M., Gross, C., Surget, S., Battaglia, F., Dulawa, S., Weisstaub, N.,

Saucy, R. (2013). *Minding the heart: The way of spiritual transformation*. Grand Rapids, MI: Kregel.

Schwartz, J., & Gladding, R. (2012). *Your are not your brain*. New York, NY: Avery.

Shapiro, S., Schwartz, G., & Bonner, G. (1998). Effects of mindfulness-based stress reduction on medical and premedical students. *Journal of Behavioral Medicine, 21*(6), 581–599.

Sheline, Y. I. (2000). MRI studies of neuroanatomic changes in unipolar major depression: the role of stress and medical comorbidity. *Biological Psychiatry, 48*(8), 791–800.

Siegel, D. (2012a). *The Developing Mind,* (2nd ed.). New York, NY: The Guilford Press.

Siegel, D. (2012b). *Pocket guide to interpersonal neurobiology*. New York, NY: Norton & Company.

Slagter, H., Davidson, R., & Lutz, A. (2011). Mental training as a tool in the neuroscientific study of brain and cognitive plasticity. *Frontiers in Human Neuroscience, 5*(17), 1–12.

Squire, L. R. (1992). Memory and the hippocampus: A synthesis from findings with rats, monkeys, and humans. *Psychological Review, 99*(2), 195–231.

Thera, N. (1962). *The heart of Buddhist meditation: A handbook of mental training based on the Buddha's way of mindfulness*. Mawatha Kandy, Sri Lanka: Buddhist Publication Society.

Trousselard, M., Steiler, D., Claverie, D., & Canini, F. (2014). The history of mindfulness put to the test of current scientific data: Unresolved questions. *L'Encephale, 40,* 474–480.

Willard, D. (2002). *Renovation of the heart: Putting on the character of Christ*. Colorado Springs, CO: NavPress.

SUGGESTIONS FOR FURTHER READING

Saucy, R. (2013). *Minding the heart: The way of spiritual transformation*. Grand Rapids, MI: Kregel.

Moreland, J. (1997). *Love your God with all your mind: The role of reason in the life of the soul*. Colorado Springs, CO: Navpress.

Willard, D. (2002). *Renovation of the heart: Putting on the character of Christ*. Colorado Springs, CO: NavPress.
Brother Lawrence. (1982). *The practice of the presence of God*. New Kensington, PA: Whitaker House.
Siegel, D. (2012a). *The developing mind* (2nd ed). New York, NY: The Guilford Press.
Newberg, A., & Waldman, M. (2009). *How God changes your brain*. New York, NY: Ballantine Books.
Schwartz, J., & Gladding, R. (2012). *Your are not your brain*. New York, NY: Avery.

CHAPTER 6

WE WERE MADE FOR THIS

Reflections on the Mirror Neuron System and Intercultural Christian Education

Timothy Paul Westbrook
Harding University

Although the intercultural classroom is not a new phenomenon, it is the reality teachers of higher education face in the 21st century. When teachers, who were trained in a monocultural system, encounter ethnic and international diversity, they may experience a bit of culture shock. If this is true for the teachers, one would expect students likewise to experience cross-cultural disequilibration. Culturally aware teachers in such settings must extend a great deal of effort in exploring the variety of expectations, fears, epistemological frameworks, and motivators that their students bring to the learning process. Though cultural differences could create barriers to learning, variegated classrooms may also result in robust learning experiences. Christian educators, then, have an opportunity to transform the challenges of cultural differences into resources for finding meaning and for developing spiritual values (Westbrook, 2015b).

Intercultural classrooms, whether the cultures originate from within one nation or multiple nations, can challenge Christian educators to consider the theological implications of cultural diversity. Genesis 1:26–28 recounts how God created humankind in his image and how he set in motion a framework for understanding theology more completely through human relationships. According to this framework, God works in *community* and his work is seen through *human activities*. God appointed humankind *to govern* the created order, and God expects human governance *to respect* human dignity. Although God has created people as individuals, embedded in the nature of *imago Dei* is the tension of being autonomous people who dwell in disparate communities. This tension surfaces throughout the Bible, yet the eschatological images of the throne room of God present hope for unity in diversity as a multiethnic chorus sings honor and praise to God and to the Lamb (Rev. 5:9, 7:9–10). It is these primordial and eschatological brackets that warrant Christian education to work toward a unified, multicultural learning community within institutions of higher learning.

Furthermore, social science studies in brain-based learning offer insight into how the neurological systems process interpersonal behavior. In particular, research in the human mirror neuron system (hereafter MNS) demonstrates an intriguing mental mimicking that occurs when people see, hear, or think about motor movements and emotions. As a natural biological feature, the MNS could be utilized as a tool for learning and close the gap between the culturally familiar and unfamiliar. In short, an understanding of the MNS assists educators to think creatively about designing learning activities and educational systems for intercultural learners.

This chapter explores how neuroscience informs intercultural Christian education, judging cultural diversity to be a resource for Christian higher education. In order to accomplish this goal, mirror neuron research and Laurent Daloz's transformational learning model for engaging otherness are discussed below through a theological lens in such a way that assists Christian educators in higher learning to consider new ways for the cultural backgrounds of students to be tools for learning and spiritual blessings for students and teachers alike.

MACAQUE SEE, MACAQUE DO: THE MIRROR NEURON SYSTEM

In a blog posted to the Edge.org, Ramachandran (2000) made the bod statement that "the discovery of mirror neurons in the frontal lobes of monkeys ... is the single most important 'unreported' story of the decade." The course of time will evaluate the accuracy of Ramachandran's prediction, but the mirror neuron phenomenon has unleashed an active decade of studies to determine the homologous regions in the human brain comparable to the mirror neurons in the premotor cortex and pari-

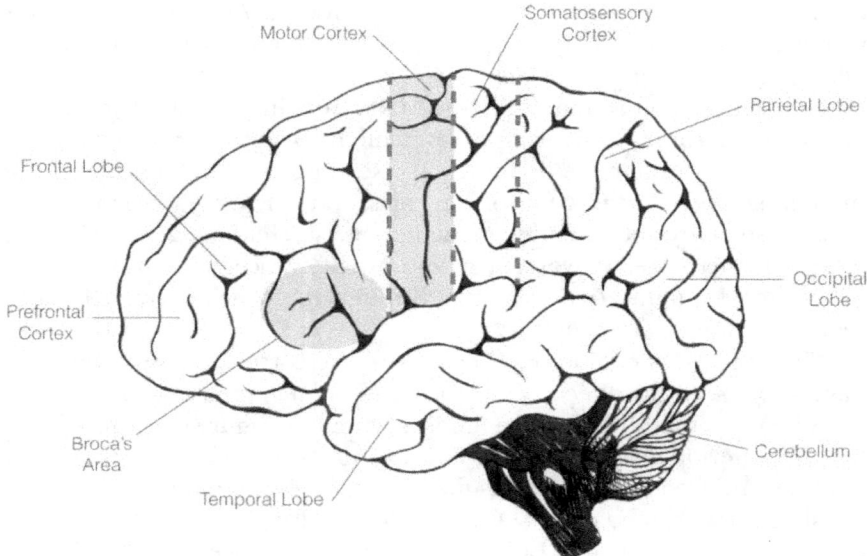

Figure 6.1. The human brain. The MNS begins with transmission in the motor and somatosensory cortices and also the Broca's area. Illustration by Megan Giddens.

etal lobules of the macaque monkey. (For a basic diagram of the human brain, see Figure 6.1.) Mirror neuron studies have been applied in a variety of research areas within the social sciences, such as human development, empathy, observational learning, understanding the "intentions of others," social cognition and imitation, language and culture development, and autism.

MNS research originated in the early 1990s when a team of scientists in Parma, Italy, discovered that in macaque monkeys the same neurons responsible for goal-directed movements (F5 region) also activated when a monkey observed lab workers' motor activities without any "overt movement of the monkey" (di Pellegrino, Fadiga, Fogassi, Gallese, & Rizzolatti, 1992). In short, the scientists deciphered that when monkeys merely observed an animal or human engaged in a transitive action, the F5 region imitated mental processing of movement (Rizzolatti & Craighero, 2004).

As the research continued, scientists quickly turned to the human brain, seeking the existence of the MNS. Evidence was soon found for a human MNS, and researchers learned that the MNS extends into social interaction, intention to act, linguistics, nonverbal communication, and even music. The implications here are profound. Abstract associations with words and music trigger motor activity in the human brain. Merely talking about picking up a pen, for example, would activate the literal embodiment of that action within the MNS. In a similar way, young adults or teenagers simply discussing a yawn might trigger a yawn contagion momentarily throughout their group. Reading a vivid novel, viewing motion pictures, watching performing arts, and playing video games generate to some extent empathic mental imagery (Helding, 2010; Manney, 2008). Furthermore, nonverbal expressions, utterances, and gestures contribute to communication exchange through the collaborative efforts of motor and linguistic systems.

While the MNS exploration began with the cortical regions, scientists have also traced the effects of mirroring on human emotions via connections between the MNS and the limbic system (see Figure 6.2). In particular, the anterior insula, an inner cortical substance below the temporal, parietal, and frontal cortex, joins the motor cerebral regions with the limbic system (Molnar-Szakacs & Overy, 2006; Nelson et al., 2010; V. Ramachandra, Depalma, & Lisiewski, 2009). Evidence of the MNS's association with the limbic system could be seen as emotional stimuli instigating changes the heart rate and body temperature. While results varied in the degree of impact made by hearing emotional sounds and thinking about emotions, the findings suggested amygdalic mirroring of both auditory sensing and cognitive reflection on emotions.

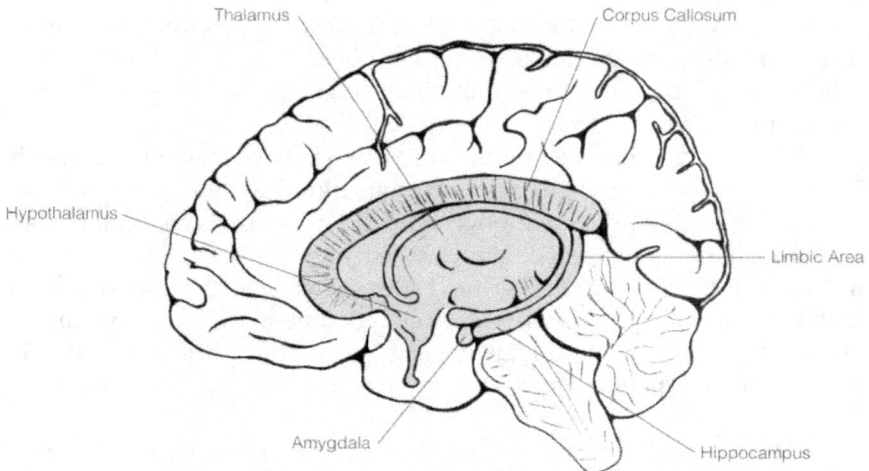

Figure 6.2. The limbic system. The MNS results in an affective response through the limbic system. Illustration by Megan Giddens.

A compelling example of the MNS's affecting the limbic system would be the study conducted by Pfeifer and colleagues (2008) that, using functional magnetic resonance imaging (fMRI), examined the connection the MNS has with empathy and interpersonal competence among children. During the experiment, the children were shown slides of emotional facial expressions. With some slides the children were asked simply to observe the faces, but with other slides the children were asked to imitate the expression. The findings showed that both observation and imitation activated the "MNS-insula-amygdala circuit" (Pfeifer et al., 2008, p. 2079).

Depending on the type of stimulus, or perhaps combination of stimuli, when people view, hear, or participate in action-oriented activity, the occipital, temporal, and parietal lobes work in conjunction with the premotor cortex (Biermann-Ruben et al., 2008). Once mirroring begins, neurotransmission travels through the anterior insula into the limbic system. When emotional mirroring is involved, the amygdala mirrors the effect, sending signals to the observer or imitator to experience the reflected emotional response (Pfeifer et al., 2008).

DO AS I SAY AND AS I DO: CROSSING THE SYNAPSE BETWEEN THE MIRROR NEURON SYSTEM AND INTERCULTURAL EDUCATION

MNS research has demonstrated that when one presents known stimuli a mirroring effect occurs, but how does the MNS respond when the brain does not recognize stimuli? In short, the mirroring does not occur, or at least, it does not mirror in the same way. In an interesting study by Buccino and colleagues (2004), human subjects observed multiple species perform oral communication and biting movements. When people observed monkeys' and dogs' eating or lip-smacking, MNS activity resulted; but when people observed dogs' barking, no MNS activity occurred. The participants only recognized the barking visually and not as an imitable motor action. In a way, the barking dogs were producing an imperceptible stimulus to the MNS. Helding (2010) cited an experiment aired on PBS in which participants monitored by a magnetic resonance imaging (MRI) observed expert dancers. The findings showed that the MNS responded stronger to dances already known by the viewers than it did to unfamiliar dances. In a similar way, but with a different species, Immordino-Yang (2008) told of a study in which experimenters ripped paper in front of two sets of monkeys: those that had played with paper and those that had not played with paper. The MNS responded in monkeys that had played with paper. In contrast, the monkeys that had no experience with paper had no MNS response to the paper's tearing. This lack of MNS activity with unfamiliar stimuli raises important implications

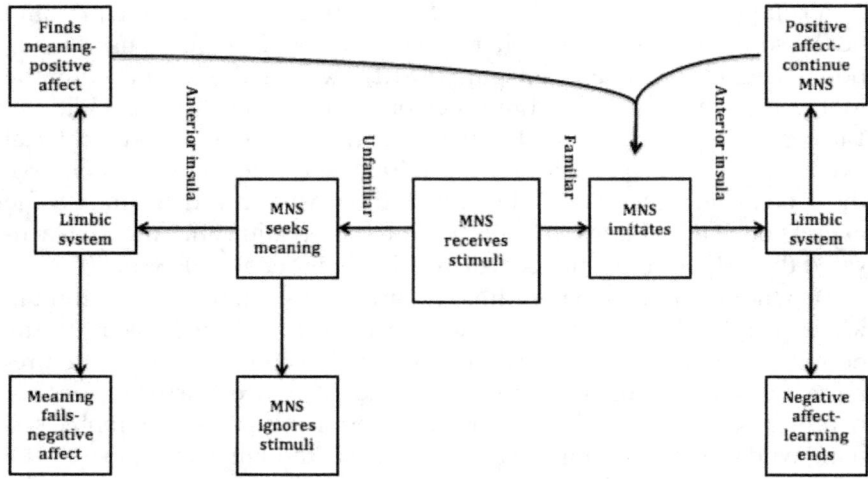

Figure 6.3. The experiential learning cycle within the mirror neuron system.

for students who find themselves in culturally different learning environments.

The MNS pathway could be seen from a learning perspective as follows. As the stimuli enter into the lobules, if the MNS recognizes the stimuli, then the neurons begin to imitate the stimuli at the motor level. As the neurotransmitters continue through the anterior insula into the limbic system, if the stimuli produce negative affect, then most likely, the MNS will cease or the individual will look for ways to end the negative affective response. If the stimuli produce positive affect, then the MNS will likely continue, and the individual finds neurological peace and harmony between the MNS and the limbic system (see Figure 6.3).

When one ponders the interaction between experience, perception, and neurological activity, it is difficult to miss the similarities with Dewey's (1938, pp. 42–43) learning situation, where human interaction and continuity of experience cross paths. Kolb's (1984) axes of prehension (thinking/acting) and transformation (reflecting/experimenting) in his Experiential Learning Theory seem to describe nicely the interplay of the MNS and cognitive development. Indeed, even thoughts of learning-by-doing philosophers such as Rousseau, Pestalozzi, and Froebel come to mind when one views the image of neurological imitation of motor activity with a purpose (Froebel, 1906; Pestalozzi, 1898; Rousseau, 1979). The convenient parallels between educational philosophy and neuroscience help educators see anatomical functionality as material evidence of previously noted processes of human behavior.

If the brain receives stimuli that are unfamiliar or uninteresting to the MNS (i.e., the MNS does not recognize it as an action or motor activity), the MNS might ignore the stimuli, as in the case of human's perceiving the barking dog or the monkey's experience with paper. Yet the human brain seeks meaning, and the MNS might try to imitate the unfamiliar stimuli based on whatever familiarity it recognizes, as in the case of the dancers who have weaker MNS responses to unfamiliar dances than to familiar dances. This may be the cellular level version of disequilibration, or to use Mezirow's terms, a disorienting dilemma (Mezirow, 1991). When a person perceives limited familiarity, the meaning-seeking neurons signal to the limbic system, which leads to either a positive or negative affective response to the new stimuli. Of course, negative affect hinders learning, but a positive response continues the meaning seeking and reinforces the brain to continue to find familiar aspects of the stimuli to imitate. Developmental psychologists might label this part of the mirroring process "assimilation" (see Figure 6.3).

As one considers the difficulties of adjusting to a new culture, the relationship between the MNS and unfamiliarity becomes apparent. While the international students' brains search for meaning in their unfamiliar contexts, their MNSs might miss social, emotional, or linguistic cues. The MNS does not know what it does not know; rather it only imitates stimuli that it recognizes. Failure to engage in MNS responses during learning situations not only could have a negative effect on students' grades, but it could also lead to embarrassment, give the appearance of apathy, and result in negative affect toward the subject being learned or toward the teacher. A goal, then, for teachers with intercultural classrooms would be to look for ways to engage the students based on their own cognitive and motor resources in such a way that creates familiarity and crosses cultural gaps. This takes a concerted effort on behalf of the teachers to be familiar with students in the classroom, to be aware of their responses or lack of responses in the learning environment, and to be creative in how they present course content, assignments, and assessments.

AN IMITABLE MODEL FOR INTERCULTURAL EDUCATION

Christian higher education faces the challenge of learning to live in community with people of differing ethnicities, racial locations, cultural values, religious beliefs, ideological positions, and socioeconomic levels. As educational institutions become more diverse, more attention should be given for critical evaluation of what are considered social norms and status quo. Predominantly white schools or European and American centered educational systems must address the extent to which race and ethnicity affect policies and expectations on what is regarded as normative. In response, Christian educators ought to reflect theologically on the implications of diversity. For example, in the Tower of Babel narrative (Gen 11:1–9), was God punishing hubris, diversifying humanity as an extension of his creation, or some combination of both? In the parable of the Good Samaritan, was it not the despised Samaritan who was more of a neighbor than the Jewish elite? This story exemplifies a more inclusive theology of embracing ethnic otherness (Luke 10:29–37). Perhaps Pentecost serves not as a reversal of the "curse of Babel," but as a demonstration of the power of the Spirit and the intention of God to create oneness that transcends diverse people (Acts 2).

Whereas the dispersion in Genesis 11 aligned with God's *tertio creatia*, the advent of Christ marks a new era in which *imago Dei* and defining the people of God not genealogically, but Christologically (Col 1:15–20). Peace and reconciliation for humankind are established through Christ (Westbrook, 2015a, pp. 65–66). Indeed, the image of Revelation 7:9–10 of people "from every nation, from all tribes and peoples and languages" reminds Christians that God's people are not homogenous and when applying this vision as well as the teachings of Jesus to Christian education, the New Testament provides theological reasons for educators to embrace cultural diversity in learning. Just as the throne of God will be surrounded by a multilingual, multicultural chorus, students and teachers ought to consider how diversity in Christian education might enhance how people experience the world as well as how they reflect on the one who made it.

Transformational learning theorist, Laurent Daloz (2000), presented four "conditions" needed to facilitate "engagement with otherness": "presence of the other," "reflective discourse," "a mentoring community," and "opportunities for committed action." Daloz's model fits well with this current exploration of neuroscience, theology, and intercultural engagement, since the model's categories assist educators to focus on how to draw from "otherness" as a resource. As people with varying backgrounds interact, the goal, then, would be for the interaction to generate new neurological pathways that lead to appreciation of diversity, to inspire new experiences that inform theological reflection, and to treat cultural others respectfully for the glory and honor of God.

PRESENCE OF THE OTHER

The first and more obvious condition is that ethnic "others" must share common experiences for cultural barriers to be crossed. Gardner (2008, p. 132) correctly opined, "Every time we are exposed to a new individual—in person or in spirit—our own horizons broaden." According to Daloz, social strength grows out of diversity, but for diversity to lead to constructive development in a society, otherness must be introduced in manageable doses. In MNS terms, too much unfamiliarity can cause the brain to reject or ignore the unexpected differences; but given enough repeated common experiences and the creation of a context, the brain creates new neurological pathways and learns to accept and even imitate the actions of cultural otherness (Del Giudice, Manera, & Keysers, 2009). In theological terms, the cross-cultural interaction influences how people view cultural others from a spiritual frame of reference. Not only do positive intercultural experiences set in motion positive relationships between cultural others, but these positive experiences would lead to a greater interest in affirming the image of God in humankind.

Teachers with students from a variety of cultural backgrounds have built into the class dynamics important resources for broadening cultural horizons. Instead of expecting the international students to learn to blend in, one strategy would be to invite them to share about their cultures in an open way for the benefit of the whole class. While these students discuss their own heritages, customs, and beliefs, the teacher could draw analogies from the similarities between the students' comments and desired learning outcomes for the course.

Another strategy would be to group students interculturally and assign group tasks. While the students spend time working together on a common problem, they will naturally wrestle with cultural barriers that they must overcome in order to complete their projects successfully. In the meantime, the teacher would be observing what issues arise and would be available to help students work through social or cultural impasses. Cultural exposure in a safe and controlled environment invites positive affect to reinforce the neurological meaning making.

REFLECTIVE DISCOURSE

A second important element to Daloz's social transformation model entails reflective, critical dialog with cultural others. Mezirow defined reflective discourse as "specialized use of dialogue devoted to searching for a common understanding and assessment of the justification of an interpretation or belief" (2000, pp. 10–11). This condition takes a step further than simply sharing common space with cultural others. Reflective discourse implies sharing personal narratives, beliefs, and impressions for the sake of attaining deeper understanding between two or more people. As students dialog and build bridges between common values, they learn new categories for theological reflection. This exchange occurs while students exhibit empathy, which is a biological function within the

MNS. When students listen to each other's stories and words, the MNSs mirror the emotional cues exchanged in dialog. Perhaps the students would even start synchronizing in posture or hand gestures (Carr, Iacoboni, Dubeau, Mazziotta, & Lenzi, 2005). While the nonverbal ballet commences, students find commonalities onto which they can restructure their worldviews but maintain their cultural distinctions.

One way for students to share in reflective discourse would be for them to divide into cultural groups but to be given a course related value that would be shared by all. For example, in an Old Testament course there may be a unit that teaches about family systems. The teacher might have students compare the similarities between filial piety of Confucian Heritage Cultures and family dynamics in North America, and discuss what it means to honor one's parents (Exodus 20:12) (Westbrook, 2012). Divided into cultural groups, the Confucian Heritage Culture students could report what filial piety means in their culture, and North American students could share what family values means to their culture. The student reporting could be in the form of discussion, oral reporting, demonstration of an expressive art piece, performance of a song, or any other medium that dips into the learning preferences of the students as well as demonstrates cultural distinctives. After the cultural exchange, the group could discuss the biblical virtue of honoring father and mother and how that might look in different cultures. Such sharing would achieve course requirements, enhance cultural awareness, and stimulate theological reflection.

Another form of reflective discourse would be intrapersonal discourse and reflection. The teacher could assign reflective essays in which students would try to think through the cultural others' experiences of life. Students could also reflect on their own feelings and reactions to being in an intercultural environment. As the teacher reviews the students' essays, she or he would look for indicators of openness, curiosity, and acceptance as well as indicators of disdain, confusion, and frustration. Not only do such assignments generate metacognition for the students about their own biases and feelings about intercultural engagement, the feedback from students helps guide the path of the teacher's lesson plans in terms of gauging appropriate levels of disequilibration that give birth to new ways of viewing the world. If such reflection remains positive and rewarding for the students, the students are more likely to enter into common areas with cultural others and have their horizons broadened.

A MENTORING COMMUNITY

Daloz suggests that a "mentoring community" can help people process cultural otherness in positive ways. He wrote, "we have found that people who are able to act on and sustain long-term commitment to a positive vision often describe in their past or current lives a 'mentoring community'—an ecology of relationships with people who value diversity and transformative discourse" (Daloz, 2000, p. 116). To establish a mentoring community for intercultural college students implies the need for teachers to seek informal relationships and encourage a safe environment in which mentoring relationships may be formed. Teachers who are already pressed for time for grading, for preparing lectures, for serving on institutional leadership positions, and for fitting in research may resist the notion of adding educative moments with students out of the classroom. However, if teachers might reflect a bit on the moments of major impact from their own favorite professors, they might recall that just a half hour or less of meaningful conversation resulted in life-altering moments. University professors should be willing to look for cultural others in their classrooms and creatively explore ways to be a source of encouragement to their students before class, after class, in the halls, over coffee, and even over a meal.

In addition to teacher-student mentoring, student-to-student mentoring could be considered. Some students who come from a high power distance culture might regard efforts by the professors to socialize as suspicious and unorthodox. Where students resist entering into mentoring relationships with teachers, perhaps the newcomers to the university might be more accustomed to befriending other students. For this to occur, faculty or staff, such as student life directors as well as teachers, could select students to train for becoming mentors. Whether teacher-student or student-student, the social interaction, conversations, and common experiences create cultural bridges that allow the MNS to function in spite of cultural differences.

OPPORTUNITIES FOR COMMITTED ACTION

Daloz's final condition appropriately calls for people interested in social transformation to accept a disposition of commitment to change. If we may draw, again, from the narrative of the Tower of Babel, this story reflects the difficulties people have in living in harmonious communities where languages differ. Though this human tendency was tapped for the purpose of scattering humanity, later in Scripture God reunites the diverse populace through the advent of his Spirit (Acts 2:1–12). As noted above, when the Pentecost crowd heard the apostles in each of their languages, this divine act was not a reversal of Babel but rather a sign that what God once separated was now brought together in Christ. God signaled unity in spite of diversity at Pentecost. Daloz correctly calls for a commitment to change, and educators who are Christian may seize globalization as a way to demonstrate Christian unity in diversity.

For the sake of bolstering diversity and intercultural engagement as a Christian responsibility with educational benefits, this chapter offers the following recommendations.

- *First, action begins at the point of entry for students.* From admissions, the business office, to records, supportive staff set the first impression for incoming students. Not only should the staff receive training in working cross-culturally, but leaders in these offices could designate one or more people who could be contacts for students from other cultures, people who are equipped with skills for shepherding new students through low context, impersonal processes.
- *Second, the student life office should be equipped to respond to intercultural issues that arise in such a way that creates a supportive community.* Support networks are key to successful student completion for underrepresented students (Westbrook, 2015a, pp. 169–170). If no one in the student life office is familiar with intercultural dynamics, existing personnel could receive training, but the office might also consider hiring new personnel who could advocate for intercultural students on campus.
- *Third, administrators could organize faculty development meetings and seminars that raise awareness of problems intercultural students face.* These faculty development opportunities could also suggest best practices for teachers who have international or otherwise ethnically diverse student rosters.
- *Fourth, academic leaders must lead the charge for committing to fair learning environment through mission statements, institutional review outcomes, public relations campaigns, hiring strategies, and through creating new budgetary line items that secure funding for institutional movement toward academic access and success for all.* A strong commitment to inclusiveness by the school's leadership must be present for institutional culture to change.

CONCLUSION

The interculturalization of the American classroom is not a dream for the future but a present reality. American universities ill equipped to embrace cultural diversity miss many opportunities for creating learning environments with robust cultural exchange and deep learning. Mirror neuron research suggests that humans naturally imitate actions and emotions.

The research also demonstrates that when new stimuli are perceived, unless one intentionally accommodates that newness into one's ways of thinking and acting, the tendency is to resist the change and remain egocentric and ethnocentric. However, the human brain is designed to search for meaning out of meaninglessness; and intercultural engagement, although disorienting at times, stimulates learning and broadens people's worldviews. Through the complex neurological process, humans communicate, exchange behavioral patterns, learn new ways of seeing the world, and acquire new values. The MNS research helps Christian educators gain biological insight into how to stimulate new knowledge, instigate equity for the classroom, and guide a global community of learners.

Christian higher education bears a theological responsibility to create learning environments in alignment with the mission of God. Though neuroscience enlightens the educator on biological processes that influence social interaction, the theological thread for peace and reconciliation in Christ reminds Christian higher education that intercultural classrooms present institutions with an opportunity to guide people from multiple cultures and perspectives to bring glory and honor to God. Not only would God be praised, but students and teachers alike would have their theological horizons broadened as they exchange their various experiences of God.

DISCUSSION QUESTIONS

1. In what ways have you seen nonverbal behavior imitated in social settings?
2. How would you see mirror neuron research influencing educational strategies?
3. What would you consider the most important reasons for cultural others to share the same educational spaces?
4. If there have been times when you experienced cultural or racial underrepresentation, what kinds of emotions did you have?
5. If it is true that God designed diversity, then what does this suggestion teach us about the nature of God?

REFERENCES

Biermann-Ruben, K., Kessler, K., Jonas, M., Siebner, H. R., Bäumer, T., Münchau, A., & Schnitzler, A. (2008). Right hemisphere contributions to imitation tasks. *European Journal of Neuroscience, 27,* 1843–1855.

Buccino, G., Lui, F., Canessa, N., Patteri, I., Lagravinese, G., & Benuzzi, F. (2004). Neural circuits involved in the recognition of actions performed by nonconspecifics: An FMRI study. *Journal of Cognitive Neuroscience, 16*(1), 114–126.

Carr, L., Iacoboni, M., Dubeau, M. C., Mazziotta, J. C., & Lenzi, G. L. (2005). Neural mechanisms of empathy in humans: A relay from neural systems for imitation to limbic areas. In J. T. Cacioppo & G. G. Berntson (Eds.), *Social neuroscience: Key readings* (pp. 143–152). New York, NY: Psychology Press.

Daloz, L. A. P. (2000). Transformative learning for the common good. In Jack Mezirow & Associates (Eds.), *Learning as transformation: Critical perspectives on a theory in progress* (pp. 103–123). San Francisco, CA: Jossey-Bass.

Del Giudice, M., Manera, V., & Keysers, C. (2009). Programmed to learn? The ontogeny of mirror neurons. *Developmental Science, 12*(2), 350–363.

Dewey, J. (1938). *Experience and education.* New York, NY: Touchstone.

di Pellegrino, G., Fadiga, L., Fogassi, L., Gallese, V., & Rizzolatti, G. (1992). Understanding motor events: A neurophysiological study. *Experimental Brain Research, 91*, 176–180.

Froebel, F. (1906). *The education of man* (W. N. Hailmann, Trans.). New York, NY: D. Appleton.

Gardner, H. (2008). Who owns intelligence? *The Jossey-Bass reader on the brain and learning* (pp. 120–132). San Francisco, CA: Jossey-Bass.

Helding, L. (2010). The mind's mirrors. *Journal of Singing, 66*(5), 585–589.

Immordino-Yang, M. H. (2008). The smoke around mirror neurons: Goals as sociocultural and emotional organizers of perception and action in learning. *Mind, Brain, and Education, 2*(2), 67–73.

Kolb, D. A. (1984). *Experiential learning: Experience as the source of learning and development.* Upper Saddle River, NJ: Prentice Hall.

Manney, P. J. (2008). Empathy in the time of technology: How storytelling is the key to empathy. *Journal of Evolution and Technology, 19*(1), 1–11.

Mezirow, J. (1991). *Transformative dimensions of adult learning.* San Francisco, CA: Jossey-Bass.

Mezirow, J., & Associates. (2000). *Learning as transformation: Critical perspectives on a theory in progress.* San Francisco, CA: Jossey-Bass.

Molnar-Szakacs, I., & Overy, K. (2006). Music and mirror neurons: From motion to 'e'motion. *Social Cognitive and Affective Neuroscience, 1*(3), 235–241.

Nelson, S. M., Dosenbach, N. U. F., Cohen, A. L., Wheeler, M. E., Schlaggar, B. L., & Petersen, S. E. (2010). Role of the anterior insula in task-level control and focal attention. *Brain Structure and Function, 214*(5), 669–680.

Pestalozzi, J. H. (1898). *How Gertrude teaches her children* (E. Cooke, Ed., L. E. Holland & F. C. Turner, Trans.). Syracuse, NY: C.W. Bardeen.

Pfeifer, J. H., Iacoboni, M., Mazziotta, J. C., & Dapretto, M. (2008). Mirroring others' emotions relates to empathy and interpersonal competence in children. *NeuroImage, 39*(4), 2076–2085.

Ramachandra, V. (2000, May 31). *Mirror neurons and imitation learning as the driving force behind the great leap forward in human evolution.* Retrieved from http://www.edge.org/conversation/mirror-neurons-and-imitation-learning-as-the-driving-force-behind-the-great-leap-forward-in-human-evolution

Ramachandra, V., Depalma, N., & Lisiewski, S. (2009). The role of mirror neurons in processing vocal emotions: Evidence from psychophysiological data. *The International Journal of Neuroscience, 119*(5), 681–690.

Rizzolatti, G., & Craighero, L. (2004). The mirror-neuron system. *Annual Review of Neuroscience, 27*(1), 169–C–4.

Rousseau, J. J. (1979). *Emile or on education* (A. Bloom, Trans.). New York, NY: Basic Books.

Westbrook, T. P. (2012). An investigation into the effects of Confucian filial piety in the intercultural Christian education experience. *Journal for the Study of Religions & Ideologies, 11*(33), 137–163.

Westbrook, T. P. (2015a). *"Breaking the cycle": A phenomenological inquiry into the perceptions of African American adult learners at faith-based, predominantly white adult degree completion programs* (Doctoral dissertation). Trinity Evangelical Divinity School, Deerfield, IL; Ann Arbor: ProQuest/UMI.

Westbrook, T. P. (2015b). New reflections on mirror neuron research, the Tower of Babel, and intercultural education. *Christian Higher Education, 14*(5), 322–337. Chapter adapted from this publication, used with permission.

SUGGESTIONS FOR FURTHER READING

Fischer, K., and Immordino-Yang, M. H. (2008). *The Jossey-Bass reader on brain and learning.* San Francisco, CA: Jossey-Bass.

Goleman, D. (2006). *Social intelligence: The new science of human relationships.* New York, NY: Bantam Books.

Merriam, S. B., & Associates. (2007). *Non-Western perspectives on learning and knowing.* Malabar, FL: Krieger.

Mezirow, J., & Associates. (2000). *Learning as transformation: Critical perspectives on a theory in progress.* San Francisco: Jossey-Bass.

Slethaugh, G. E. (2007). *Teaching abroad: International education and the cross-cultural classroom.* Hong Kong, China: Hong Kong University Press.

Sousa, D. A. (2011). *How the brain learns* (4th ed.). Thousand Oaks, CA: SAGE.

CHAPTER 7

CHRISTIAN EDUCATION AS EMBODIED AND EMBEDDED VIRTUE FORMATION

Brad D. Strawn and Warren S. Brown
Fuller Theological Seminary

It could easily be argued that the average college or graduate student (or professor for that matter) conceptualizes education as the dissemination and acquisition of information. The idea goes something like this: students should attend the most prestigious academic institutions they can in order to obtain the greatest, latest, and most accurate information. If the student can obtain a significant amount of the right kind of information, then they will have been educated, which will lead them to becoming productive members of society.

In this chapter we will first explore a model of human anthropology that supports this understanding of education—mind/body dualism. We will then suggest an alternative anthropology that is referred to as embodied and emergent, including some of the cognitive and neuropsychology that supports it. Next we will explore how this alternative model fits nicely with James Smith's understanding of education as liturgical formation and explore how this approach may lead, not just to information acquisition, but virtue formation. Lastly, we will discuss some practical pedagogical concerns to create the kind of practices necessary for this kind of formation to flourish.

DUALISTS MODELS OF PERSONS

The model of education as *information acquisition* is actually based on a long-standing view of the human person—body/mind or body/soul dualism. This model has its roots in Greek philosophy as well as the writings of St. Augustine and René Descartes. Simply stated, this model of human beings conceives of persons as made up of two separate parts: a body and soul/mind. Augustine's work, which has been highly influential in western branches of Christianity, was based on Platonic dualism of separating body from soul, which led Augustine to the idea of an inner self, as well as an inward conceptualization of spirituality. Descartes conceptualized the body as a machine, and the mind (or the rational soul) as immaterial. He also posited that the soul was hierarchically superior to the body in which the soul resided (Murphy, 2006).

While much ink has been spilled debating this view, for the purposes of this chapter we are interested in the *potential implications* of this model for education and moral formation. The potential implications of this dualist model are multiple, but some of the most important ones for our purposes include: (1) the most important part of the person is an inner nonmaterial soul/mind; (2) the body is presumed to be less important and unreliable; (3) the body is distanced and subordinate to the soul/mind; (4) all the really important events—all that is essential about the person—are inside; (5) our obligation as Christians is inward not outward;(6) since the hierarchically privileged inner soul/mind is private, persons are isolated and essentially independent of their outward behavior and community; and (7) in the end we become committed to inward privatized spirituality, and thus to individuality.

This dualist philosophical and theological model of human persons is mirrored in much of the cognitive psychology literature, which understands human cognition (thinking) as *information processing* conceptualized via computer metaphors. In this modern materialist model of dualism, the human brain is akin to the hardware while the mind is the functioning of software (Daniel & Dennett, 1991). Input enters the brain via the body and is worked on by the brain/mind which leads to output. So we might say educationally that "good information in leads to good outputs," or the opposite, "bad information in leads to bad outputs." If this is an accurate picture of human personhood and how we learn, then it makes sense that the emphasis in education should be primarily on the information that one is acquiring.

The metaphor of human person as computer is powerful and *partially* true. Information and knowledge that one obtains does in fact influence thinking/feeling and behavior, although perhaps not in quite the ways we have imagined. Nevertheless, the information processing view is a dualist model based on an understanding of persons as made up of two things or parts: either a body and an immaterial soul/mind, or a body and inner abstract brain information processing. From this vantage point, the real action of thinking, or cognition, is the "inside work" of the mind/brain, whereas the body is not much more than a means to deliver input to the brain. All the important events—thinking, learning, and spirituality—are inside the person, which leads to inwardness, individuality and a disregard, if not degradation of the body (and potentially other bodies). Learning is something one does inside one's brain/mind with bodily events only providing secondary scaffolding. And because this model also fuels modern education, the acquisition of disembodied information, the question becomes, "What is the most expedient and cost-effective way to deliver information?" The answer is obvious: any platform that can deliver information in the fastest and cheapest manner. Students don't need to be physically present in class or engage with professors or other students – all one really needs is a means to deliver and acquire information. And lastly, we suggest that this dualist model of personhood also dominates the literature on Christian education, at least in Western ecclesiastical circles. Believers are encouraged to get the right kind of information (in this case belief) and sanctified living will result.

EMERGENT MODELS OF PERSONHOOD

In recent years Christian alternatives have risen to this dualist position, which understand persons as emerging from, but never separate from, embodied and socially embedded life. There are several variations of this model including emergent monism, nonreductive physicalism, dual-aspect monism, and others (Brown, Murphy, & Maloney, 1998: Brown & Strawn, 2012; Green, 2004; Green & Palmer, 2005; Jeeves, 2004; Jeeves & Brown, 2009). What these models share, in one form or another, is that human personhood arises as an emergent property of a hyper complex organism (brain and body) as it interacts within a physical and social environment. Therefore, concepts such as "mind," "soul," "spirit," "self," and "person" describe *aspects* of whole persons. Even in emergent models that posit an immaterial "soul," the soul is so inextricably connected to the body that they essentially function as a unity (Hasker, 2005). These emergent models are important because they suggest that persons are both embodied—including the body as well as the brain—and embedded—

meaning always contextualized within action within the world. *Contextualized action* implies that it is impossible to accurately describe or account for a person outside of her context.

An important version of emergentism within psychology and philosophy of mind is *embodied cognition* (Clark, 1998; Johnson 2007; Lakoff & Johnson, 1999). Embodied cognition challenges the information processing model by arguing that mind comes about through bodily interactions with the world. Therefore, the nature of the body and its actions in the world influence the nature of the mind. A fanciful thought experiment illustrates this. If one had the body of an elephant and the brain of a human, one would have a very different mind. So, as an extreme example, if one then learned to play basketball as an elephant, one would have a very different understanding of basketball than a being with a human body.

Embodied cognition argues that mental process are founded and grounded in actions of the body, which means that cognition is rooted in sensory-motor interactions of the body with the world. All aspects of "minding," including learning, are shaped by the nature of the body, and cognitive processing remains rooted in the actions of the body in the physical world. We think and learn, not by noting and manipulating abstract concepts (i.e., information processing), but by interacting with the world in and through our bodies. Even abstract concepts can be understood as being learned through embodied experience. For example, the abstract idea of "time" is understood by metaphorical extension of bodily action—that is, time passes, it speeds up or slows down, we can go back or forward in time, and so on (Johnson, 2007).

The central principles of embodied cognition prove important to the points we are advancing in this chapter, including the ideas that cognition (thinking) is (1) for action, (2) always situated in particular contexts and influenced by those contexts, and (3) time-pressured by the pace of action. Human action (and thus cognition) is always enmeshed in recurrent situational feedback. In other words, humans remain embedded in contexts and those contexts deeply influence ongoing appraisal and action. Embodied cognition further posits that the meaning of language (semantics) roots itself in the recall of sensory-motor experiences. Furthermore, "thinking" is the off-line running of body-based simulation that does not become manifest in immediate bodily action. Finally, embodied cognition advances the idea that some of human cognition can be off-loaded into the environment. For example, we place much of our cognitive memory into smart phones, or we do better problem solving in teams where individuals can bind into a cooperative mind (Clark, 2011).

It should be clear to the reader that the embodied cognition model, based on emergent monism, is very different from Cartesian dualism or the brain/body dualism of the informational processing computer model. In the dualist model, mental life is based on the inner manipulation of abstract representations. Sensory and motor systems (body) are merely input and output buses. Thinking is the rule-bound manipulations of abstractions and can be distanced from perceiving or doing. To change one's thinking, to change belief, to learn, is to obtain new abstract ideas and to manipulate them internally in some coherent manner. This model leads to an understanding of education as primarily the acquisition of new information, and perhaps reinforcement of correct manipulations of the information.

SMITH'S VISION OF CHRISTIAN EDUCATION

Recently, philosopher James K. Smith has challenged the concept of education as knowledge acquisition, suggesting that obtaining new abstract ideas in order to shape one's "worldview" may not be the most powerful way to shape a Christian person (Smith, 2009). Smith believes that education, especially if it is to be Christian education, must be understood as *formation*. In some sense, of course, all education is formative, but his concern is that (1) students and educators often don't take the time to note what that formation is about, and (2) they do not avail themselves of what is most deeply formative.

While Smith doesn't explicitly claim an emergent perspective, his use of social psychology research and phenomenological philosophers suggests that his approach is in the embodied camp. He utilizes an embodied theological term—liturgy—to denote formational practices. Smith argues that humans are not primarily *thinking* beings or even *believing* beings, but humans are *liturgical animals* (Smith, 2009). We are formed most deeply by the embodied liturgical practices that we engage in which are provided to us by our cultural contexts. By liturgy, Smith means any cultural participatory activity that involves practices with an implicit trajectory or goal. His argues that not only do these cultural liturgies shape and form us, but these practices contain implicit images of the good life (or telos) to which they aim. Liturgies are not reserved for what happens in Christian worship, but neither can they ever be entirely secular. Liturgies are religious because they contain implicit ethical values. So Smith calls for Christian educators to exegete the cultural liturgies that they, and their students, are embedded in and the practices that they entail—whether these are the liturgy of the shopping mall, the industrial military complex, nationalism, so called "Christian Education," or the church.

EDUCATION AS VIRTUE FORMATION

Both the embodied and emergent ways of understanding humans that we began with, and Smith's ethical liturgical anthropology, assert that we become what we do. Thus, both views should provoke us to think about virtue formation and Christian education in entirely new ways – very different ways from what is implied by either body/mind or brain/body dualism.

In thinking about Christian education and formation in this light, we find ourselves most interested in *virtue theory* as epitomized by the work of philosophers such as Alasdiar MacIntyre (1999, 2007). In short, virtue theory suggests that virtuous actions arise from basic character traits of individuals. Virtue is, therefore, evident in actions that are often automatic and habitual. And these traits are linked to implicit social schemas. These schemas are learned in interaction with the world, continually manifest themselves in our actions, and form the primary content of the off-line behavioral simulations that we experience as thinking. Virtue is based on learned body-environment action schemas. It is mostly expressed in automatic actions, and less often because of deliberate decisions. Virtue is self-organizing; that is, it is based on a developmental history of physical and social interactions with the world. In Smith's language, it is the liturgies that we have participated in (consciously and unconsciously) that form our moral character and subsequent action (Smith, 2009).

Again, to summarize, Smith borrows concepts from social psychology, philosophical phenomenology, and cognitive science (embodied cognition) to argue that what is most formative in students' lives is not the information they receive, but the embodied practices in which they engage. These liturgies and their practices shape one's "heart" toward a particular telos or goal. This telos or goal has an implicit ethic and an image of the good life. And this telos is religious because it is about what the person ultimately values. If the education that we invite our students to engage in aims them toward participating in the liberal capitalistic democratic society, then their image of the good life is something akin to "whoever dies with the most toys wins." But if the liturgy of Christian education has the Kingdom of God as its telos, then we might expect that the image of the good life might be something like loving God and loving your neighbor.

Smith and contemporary cognitive science seem to agree that this doesn't happen by filling students "minds" with the right kind of information, but by inviting them into a particular kind of embodied life with its corresponding social practices. We believe (as do many others) that Christian education is formation rather than simply information acquisition.

We further believe that this formation comes about through embodied and embedded liturgical practices which lead to formation of virtues with the telos of the Kingdom of God.

MIRRORING, IMITATION, AND FORMATION

If the processes of formation are embodied and embedded, how might we understand this from the perspective of neuroscience? One resource is the recent discovery of mirror neurons. Researchers accidentally discovered that some neurons in the motor areas of a monkey's brain fire not only when the monkey was performing an activity, but also when that monkey observed someone else performing the same task. Further studies have demonstrated that similar neuronal mirroring occurs in the human brain. Such mirroring is believed to be the basis of comprehending the action of another, anticipating the action of another, speech perception, theory of mind (inferring what another individual is thinking), and even empathy (Keyser, 2011). Because our motor systems implicitly mirror the actions of others, watching others primes our behavioral systems to do the same thing (now or at some later time). Thus, such mirroring is hypothesized to be the basis of learning through imitation. Taken together, this discovery lends greater support to the idea of learning being embodied (recollections and simulations of bodily experiences—our own or the mirrored actions of others) and embedded (we are wired for imitation).

Learning and virtue formation have much more to do with what we *do* in our bodies than it does with the acquisition of bits of data or information. And what we do is strongly influenced by the persons we imitate. Thus, if our bodies are embedded in educational communities engaged in the kind of liturgical practices whose telos is the Kingdom of God, then we will be formed in virtues consistent with that aim.

PEDAGOGICAL PROCESSES

We have thus far described how deeply our bodies and their activity in the world are implicated in the nature of our minds. We have used Smith to suggest that Christian formation has much less to do with propositional beliefs and much more to do with the formative impact of the liturgies we enact (Smith, 2009). The impact of action further suggests a focus on virtues rather than knowledge as the outcome of Christian formation. Finally, we argued for the impact of imitation on formation of the mind. How might these concepts directly impact our thinking about the processes of Christian education?

First, we need to consider the impact of mirroring and imitation in Christian education and formation. The old saying is apropos here, "Don't just tell me, show me!" Professors can't employ too many examples, especially active, embodied examples that they demonstrate and embody for students. While this might be a challenge in some disciplines, it is worth the time for professors to consider what this might look like. Even examples of what not do to, or "failures," can be helpful for student formation.

Secondly, the use of stories and anecdotes can be powerful in many domains of teaching. Research suggests that a story is not disembodied information that the brain/computer inputs, works on, and spits out. Rather, a story prompts hearers to run motor simulations of the narrative in which they imagine themselves physically involved in various roles in the story, and to compare and contrast the story outcomes with stored behavioral scenarios of their own life experiences (Brown & Strawn, 2012). Daniel Siegel (2015) argues that stories seem to have an integrating function in bringing together various neural processing areas such as the limbic system and the higher cortical areas. Of course, the kind of stories would be important as well. For example, stories of exemplars that we want our students to imitate would be important, but also stories of failures can be told as cautionary tales.

This leads to a third point, the importance of creating a secure attachment in educational settings. Several different attachments styles have been identified in childhood that often persist into adulthood—secure attachment being the gold standard. Secure attachment is developed first between a child and caregiver through embodied acts of dependability which create a basic sense of safety and security in the child. Children and adults with secure attachment styles are more open and less disoriented to new experiences (Karen, 1998). This suggests that students may be better able to tolerate formational experiences that are destabilizing if they have a history of secure attachment, but also if they have a sense of secure attachment with their professor. While not everyone is lucky enough to develop secure attachment in childhood, the good news is that it can be "earned" through later relationships with others in such a way that helps one renarrate a life.

Several issues will have a profound influence on educational attachment and its impact on formation. First, presence matters. Robust learning through imitations demands embodied interactions. Second, group-size matters. At the extreme, research by Robin Dunbar (2009) has indicated that, based on primate brain-size versus group-size, humans can only manage about 150 (at most, probably less) stable and continuing relationships. Anything more (a campus of thousands, or hundreds of "friends" on Facebook) is relationally unmanageable and ephemeral.

More importantly, research in group therapy and dynamics suggests that 12 members is an optimum number for the level of group cohesion that fosters formation (Yalom, 2005). This obviously has implications for how many advisees a professor should have as well as the size of classes.

Third, time matters. A relationship that is formative must have consistent time in order to make an impact. However, as James K. Smith (2009) points out, we are embedded in cultural liturgies that are continually forming us in ways we don't even recognize. Consider the number of commercials viewed per day between print media and social media, television, billboards, and the radio. We are constantly receiving these minis "sermons" about the good life, and the practices one needs to engage in to attain this good. It is hard to imagine that meeting with a professor for 2–4 hours per week in a classroom environment will have anywhere near this level of formative impact. But education as virtue formation does not have to be limited to the classroom. Today, professors and students are finding ways to interact informally around shared meals, time in the dorms, extracurricular activities, and so on. These interactions may not be organized around the content of a course, but will undoubtedly be formative as students and professors share life together, fitting the content of the course into the contexts of life. Here, students and professors have the opportunity to engage in shared liturgies, and to learn together through mutual imitation.

CONCLUSION

The truth is that all education is formative even if it only conceives of education as information acquisition, and even if students are understood primarily through the lens of the computer model of mind. The question becomes not are we forming our students, but how are we forming them? What is the real telos of the liturgies we embody in the classroom and on campus? Is this formation embodied and embedded and is its goal virtue formation of Christlikeness? And finally, are we making the best use of what we know about human personhood to most effectively accomplish this formation?

DISCUSSION QUESTIONS

1. What are some implications of a dualist view of persons and of an emergent monist view?
2. Name some examples of what Smith calls cultural liturgies and describe their image of the good life.

3. What is so important about the ability of the brain to mirror the activity of others?
4. How is Christian education as formation different from education as information acquisition?
5. Why might we care about virtues, and how are virtues formed?
6. What are some practices that faculty and students could engage in to make learning more embodied?

REFERENCES

Brown, W. S., Murphy, N., & Malony, H. N. (Eds.). (1998). *Whatever happened to the soul? Scientific and theological portraits of human nature.* Minneapolis, MN: Fortress Press.

Brown, W. S., & Strawn, B. D. (2012). *The physical nature of Christian life: Neuroscience, psychology, & the Church.* New York, NY: Cambridge.

Brown, W. S., & Strawn, B. D. (2015). Self-organizing personhood: Complex emergent developmental linguistic relational neurophysiologicalism. In J. R. Farris & C. Taliaferro (Eds.), *The Ashgate research companion to theological anthropology* (pp. 91–102). Farnham, England: Ashgate.

Clark, A. (1998). *Being there: Putting brain, body, and world together again.* Boston, MA: MIT Press.

Clark, A. (2011). *Supersizing the mind: Embodiment, action, and cognitive extension.* Oxford, England: Oxford University Press.

Daniel, C., & Dennett, D. C. (1991). *Consciousness explained.* New York, NY: Little, Brown & Co.

Dunbar, R. (1992). Neocortex size as a constraint on group size in primates. *Journal of Human Evolution, 22*(6), 469–493.

Green, J. B. (Ed.). (2004). *What about the soul? Neuroscience and Christian anthropology.* Nashville, TN: Abingdon.

Green, J. B., & Palmer, S. L. (Eds.). (2005). *In search of the soul: Four views of the mind-body problem.* Downers Grove, IL: InterVarsity Press.

Hasker, W. (2005). On behalf of emergent dualism. In J. B. Green, & S. L. Palmer (Eds.), *In search of the soul: Four views of the mind-body problem.* Downers Grove, IL: InterVarsity Press.

Jeeves, M. (Ed.). (2004). *From cells to souls—and beyond: Changing portraits of human nature.* Grand Rapids, MI: Eerdmans.

Jeeves, M., & Brown, W. S. (2009). *Neuroscience psychology and religion: Illusions, delusions and realities about human nature.* West Conshohocken, PA: Templeton Foundation Press.

Johnson, M. (2007). *The meaning of the body: Aesthetics of human understanding.* Chicago, IL: Chicago University Press.

Karen, R. (1998). *Becoming attached: First relationships and how they shape our capacity to love.* Oxford, England: Oxford Press.

Keyser, C. (2011). *The empathic brain: How the discovery of mirror neurons changes our understanding of human nature.* Lexington, KY: Social Brain Press.
Lakoff, G., & Johnson, M. (1999). *Philosophy in the flesh: The embodied mind and its challenge to western thought.* New York, NY: Basic Books.
MacIntyre, A. (1999). *Dependent rational animals: Why human beings need the virtues.* Chicago, IL: Open Court.
MacIntyre, A. (2007). *After virtue: A study in moral theory.* South Bend, IN: Notre Dame.
Murphy, N. (2006). *Bodies and souls, or spirited bodies.* New York, NY: Cambridge.
Siegel, D. J. (2015). *The developing mind: How relationships and the brain interact to shape who we are.* New York, NY: Guilford.
Smith, J. K. A. (2009). *Desiring the kingdom: Worship, worldview, and cultural formation.* Grand Rapids, MI: Baker.
Yalom, I. (2005). *The theory and practice of group therapy* (5th ed.). New York, NY: Basic Books.

SUGGESTIONS FOR FURTHER READING

Brown, W. S., & Strawn, B. D. (2012). *The physical nature of Christian life: Neuroscience, psychology and the church.* Cambridge, England: Cambridge University Press.
Keyser, C. (2011). *The empathic brain: How the discovery of mirror neurons changes our understanding of human nature.* Kentucky: Social Brain Press.
Sian, B. (2015). *How the body know its mind: The surprising power of the physical environment to influence how you think and feel.* New York, NY: Atria.
Smith, J. K. A. (2009). *Desiring the Kingdom: Worship, worldview and cultural formation.* Grand Rapids, Michigan: Baker Books.

CHAPTER 8

NEUROSCIENCE AND CHRISTIAN WORSHIP

Practices That Change the Brain

Dean Blevins
Nazarene Theological Seminary

I appeal to you therefore, brothers and sisters, by the mercies of God, to present your bodies as a living sacrifice, holy and acceptable to God, which is your spiritual worship. Do not be conformed to this world but be transformed by the renewing of your minds, so that you may discern what is the will of God—what is good and acceptable and perfect.

(Romans 12:1–2, NRSV)

Pour out your Holy Spirit on us gathered here,
and on these gifts of bread and wine.
Make them be for us the body and blood of Christ,
that we may be for the world the body of Christ,
redeemed by his blood.

(UMH General Thanksgiving, 1989, p. 10)

INTRODUCTION

Throughout the history of Christian worship, believers gathered to celebrate God's mighty acts of creation and redemption, in the world and through Jesus Christ, in the power of God's spirit. Whether shaped by liturgical practice like the celebration of the Lord's Supper indicated in the verse above, or merely through free church practices of plain song, sermon, and prayer, believers gathered in expectation that they would be changed in and through their worship of the triune God.

From a neurological standpoint, how might that change occur? Granted, not everything that occurs within worship may be reduced to biology alone. However, if our createdness includes our inclination to worship, and if our redemption includes a transformation of mind and body as well as soul, might there not be neurological underpinnings worship's transformative power in and through the life of the congregation? This chapter explores the neurological basis for change in and through worship. Following Judy Willis (2010), the chapter begins with the "neurological" (p. 46) by finding correlates between brain activity and experiential practice. To be sure, neuroscience provides a framework for understanding how experience changes the brain by influencing how certain neurons both "fire together" and then "wire together" through the replication of experience in the worship setting. The chapter then explores how narrative and practice combine to intensify change in our brains and accompanying behaviors. Finally, the chapter explores a crucial third component within social neuroscience, one that helps Christian educators and ministers understand how worship, not merely personal devotional practices, provides a particularly powerful context for changing the brain. This final phase of the chapter begins by exploring social neuroscience, including how mirror neurons, articulates our participation in worship. The chapter closes addressing the potential of the congregation to truly provide direct influence, or causation, through both the hybridity of the mind and the place of emergent properties in congregational practice. While the final portion of this chapter may appear controversial, the combined importance of hybridity and emergence helps educators and ministers understand the "particular" place of congregational worship in shaping the brain.

Worship at the Synaptic Level

To best understand the rudiments of how experience shapes our brain we turn to the most basic building block of neuroanatomy, the synapse. Every brain cell, every neuron, exchanges information through synaptic connections. Neuroscientist Joseph LeDoux (2003) notes that we are synaptic selves, that the root of all knowledge occurs as our senses encounter experiences and then communicate the information gathered (sight, sound, smell, touch, and taste) through from one neuron's synapses to the dendrite or extensions of the next neuron. The varied experiences communicate along the body's nervous system through these connections to and through our brains, where the information is "stored" via those synaptic connections. The more information (experience) exchanged, the stronger the synaptic connections. Gradually, neurons strengthen through a process of mylenation. Myelination occurs as a fatty layer, called myelin, accumulates around the long shaft, or axon, of neurons. Myelination enables nerve cells to transmit information faster and allows for more complex brain processes through synaptic connections. As certain neurons strengthen through their synaptic connections, other neuron passages wither away through disuse. The long held Donald Hebb maxim of "neurons that fire together, wire together" undergird the idea that people find their lives, their brains, fundamentally shaped by experience (LeDoux, 2003, pp. 79–82).

At first glance, the experiential shaping of the brain through synaptic connections may seem quite deterministic, in the vein of B. F. Skinner. While the process (as we shall see) remains much more complex, neuroscience seems to provide a strong underpinning that experience shapes people. The formational quality of synaptic connections provides the beginning for understanding of how worship may shape human life. People's brains do respond to the types of experiences they encounter. In times of stress, such as in moments of fear, our bodies may respond almost before cognitive awareness (LeDoux, 2003, pp. 5–8). Like LeDoux, Daniel Goleman (2007) calls this almost autonomic response the "low road" where basic emotions (located in the limbic region of the brain closer to the brain stem) guide response before the overall "high road" of judgment centers in the prefrontal cortex (or forehead region of the brain) can activate (pp. 15–73). While the overall "wiring" of cognitive decision making includes both high and low roads, our habitual participation and engagement in worship does provide a formative effect on our brains. Athletes often discuss "muscle memory" when discussing quick, almost autonomic, response based on years of practice. Malcolm Gladwell, in his book *Blink* (2007), notes that some snap judgments, known as thin slicing, occur based on small impressions (p. 23), and then people actually often seek to justify those decisions "after" they are made based on more logical patterns (pp. 68–71).

Theologian James K. Smith (2009), in discussing worship, argues people embody *homo liturgicus*, or worshipping beings based on Augustine's notion of love, or "desires," as a combination of emotion, intellect, and will, Smith argues that worship both shapes and reflects our desires. Within a Wesleyan theological perspective, one might say that worship, as a means of grace, shapes our "affections" or "tempers" which "integrate the rationalistic and emotional dimensions of human life into holistic inclinations toward action" (Maddox, 1998, p. 40). We worship what we desire, we desire what we worship.

At the synaptic level, neuroscience provides a similar argument in what shapes our emotions and judgments through bodily actions. Neuroscientist Antonio Damasio (2010) differentiates between basic emotions, which serve primarily as a regulatory function toward external stimuli (fear when threatened), apart from social emotions that occur through interpersonal perceptions (embarrassment) and background emotions, like tension, that affect human awareness. Damasio also maintains an overall vision of the blending of body, emotion, and cognition that might concur with an Augustinian view of desire as a state that "embodies" both head and heart, cognition and emotion, as one basic process that takes place in bodies, and bodily action (Damasio, 1999; Smith, 2009).

When people gather for worship, they condition themselves through their sense experiences, their emotional regulation and memory, and in their actions in the worshipping community. The very ritual processes of worship, as bodily acts of celebration or contrition, fundamentally shape our brain processes. The stronger the experiences, the more continuous the process, the more profound the change through the strengthening of synaptic connections in the brain. Small wonder people sometimes find themselves in times of stress reacting either in fear or faith, in postures of anger or prayer. Fundamentally our participation in worship changes our brains "firing and wiring," over time, at the synaptic level.

NARRATIVE, PRACTICE, AND MEMORY

At this point of the writing, readers may note the amount of time spent in worship (compared to family life, work, or even driving a car) would not seem sufficient to create such powerful changes. Following Skinner, where behavior shapes cognition, such an assumption would appear reasonable. However, the brain does not function like Skinner's "black box" of direct input alone. As noted, synapses combine into very complex "wirings" within the brain that bring not only new sense experience to previous procedural memory, but also retrieves narrative memory as a reinforcement of the experience itself. If human beings embody *homo liturgicus*, as James K. Smith believes, they also reflect *homo narrans* (Myerhoff, 1978, p. 272), narrative and narrating beings as well.

Neuroscientists argue that narrative memory, stored throughout the brain, appears to be directly connected with the hippocampus in the aforementioned limbic region (Hogue, 2003, pp. 57–58). Long term narrative memory remains closely connected to the emotional seat of the brain, the low road, where it exerts powerful influence. David Hogue (2003) notes long term memory provides a powerful reinforcement to the working memory of a given moment. Narrative memory actually brings both our "self" as our autobiographical understanding, and our "story," our communal narrative including scripture, into play with the experience at hand (pp. 91–99).

In addition, worship often reflects additional processes of ritual practice. The very acts of worship, from greeting to dismissal, reflect particular meaning in and of themselves. To be true to Smith, true liturgy acknowledges both memory and ritual. However, for the sake of greater clarity, readers might understand ritual as a "practice" in the Aristotelian sense of an act or actions that possesses internal meaning (that holds actions together) and a social or communal understanding of excellence or virtue in those actions (MacIntyre, 2004). Worship ritual, or practice, involves actions that remain meaningful in and of themselves as we participate in them. However, these practices may also challenge a community to be more faithful, more "virtuous" in their practice. Congregational singing may be something both appreciated by a person as he or she participates, yet also acknowledged as more faithful or "spiritual" by the congregation based on a sense of presence. One might note that faithfulness might not mean the same thing as performance. Yet the "heartfelt" expression of some person either singing ... or praying ... may be acknowledged both by the participant and the congregation as truly meaningful and virtuous in his or her participation.

To return to Hogue, the combination of narrative, ritual, and experience provide a powerful triumvirate that shapes the mind (2003). To overcome the limits of sense experience alone, or Skinner's behaviorism, the mind contributes as it draws both upon the power of long term narrative memories (of the Christian story or human participation in that salvific narrative) and the meaningfulness of the practices engaged during the worshipping community. This narrative-ritual process approximates Smith's (2009) use of "social imaginary" as a comprehensive noncognitive view of the world that incorporates myth, icons, and ritual alongside narrative (p. 68). Unlike the (not so simple) task of driving a car, or completing work tasks, worship memory and practice tend to magnify the experiential connections, deepening our "self" according to those times of gathered practice.

Paul Markham (2007) develops one example that combines neuroscience, Wesley's affectional anthropology, and Christian practice as a powerful framework for personal and community formation. Markham's work remains indebted to earlier work by Nancy Murphy and Warren Brown (2007), who posit a view of humanity as embodied beings with an antireductionist or emergent consciousness. Markham combines Murphey and Brown's neuroanthropology with MacIntyre's concepts of practice and virtue to state that transformation occurs as participants engage each other and Christian practices, orienting their lives, their affections, or tempers, toward the love of God (Markham, 2007). Markham engages a large range of Christian practices, known as the means of grace, yet his theory includes aspects of worship as a formational framework that takes seriously both neuroscience and transformation. For Markham, God's activity within this process occurs at the quantum level, following the science and theology perspective of John Polkinghorne, yet requires community for true transformation. God, working at the quantum level, engages embodied human "minds" (a comprehensive term that includes embodied but emergent consciousness alongside emotion or affections) through Christian practices, transforming people toward the love of God.

Markham's approach, though limited in some perspectives (Blevins, 2009), provides Christian educators with a model that assists their appreciation for meaningful formational practices, both through relationships and personal practices. His theory, however, does not specifically address the social context of the worshipping congregation as a unique qualifier in formation. The field of social neuroscience provides additional clues that assist ministers in recognizing the unique role of worship.

ALL TOGETHER NOW: SOCIAL NEUROSCIENCE, MIRROR NEURONS, AND WORSHIP

So far in this essay, the role of meaningful narrative and practice might easily contribute to a number of formational practices, including private devotions, personal prayer, and works of service. Similarly, one might note the same level of meaningfulness in other acts of devotion, including allegiances to work, family, or sports teams. The question remains whether worship, as a corporate, social, congregational practice, offers any unique qualities for formation. The question remains crucial since people in our contemporary society rarely invest the same amount of time in worship as they do other practices or social interactions. Could some practices and interactions bear more significance than others in the formation of a person? Perhaps the field of social neuroscience, or neurosociology, bear an investigation.

Previously the fields of clinical psychology and sociology maintained quite disparate studies. Psychiatrists tended to focus on pharmacological interventions for mental conditions, while sociologists avoided biological explanations, primarily out of their rejection of social Darwinism (Schutt, 2015). As neuroscientists recognized the deep interplay between brain interactivity and social interactivity, perspectives began to change. One recent example surfaced through the writing of Antonio Damasio (2010), who concluded that cognition, emotion, and consciousness remained deeply indebted to social factors to the point that they point to a homeostatic impulse in biology and culture.

Social neuroscience remains a fairly new, interdisciplinary, field; one barely 25 years old (Schutt, Seidman, & Keshavan, 2015). Currently, researchers in the field include interest in how social interaction and brain interaction remain connected at every level (Schutt et al, 2015). Theorists working with interpersonal neurobiology, a variation within social neuroscience, provide a different disciplinary perspective between neurological processes, interpersonal relationships, social environment, and development (Siegle, 2012), as well as educational practice (Cozolino, 2013). The diverse explorations within this field allow theorists to adopt seemingly disparate approaches from personal practices in mindfulness, to interpersonal attachment theory, to sociocultural factors as alluded to above. Not all explorations prove as fruitful for Christian educators. Damasio (2010), for instance, refers to religion and myth alongside the arts as cultural constructions to maintain social homeostasis. Social neuroscientists appeal to neurological abilities to support evolutionary development through social interdependence (Schutt, 2015). Still, the field provides clear support for the acknowledged need of community for Christian formation (Brown & Strawn, 2012), as well as clues to how the worshiping community intensifies this formational process.

At the neurological level, research conducted with mirror neurons explains the powerful effect of observed emotional activity. The accidental discovery of mirror neurons revealed the power of social identification and emulation (Iacoboni, 2009). As people observe certain physical or emotional experiences expressed by other people, neurons within the observers' brains "fire" in a way that replicate the experience. People almost "participate" in the experiences of others when mirror neurons respond to certain actions or expressions (Rizzolatti & Sinigaglia, 2008). Popularized through the work of science writers such as Daniel Goleman (2011), researchers know that people literally relive certain experiences as they observe powerful, emotional experiences, of others. During worship, people not only participate in their own actions and activities, they potentially repeat the intensity and experiences of those around them. Psychologists long understood the power of group influence. Freud and Jung, unfortunately, understood those actions primarily through mob activity (Blevins, 2002, pp. 158–166). When worshippers gather, however, the embodied social power of communal liturgy, the aforementioned narrative and ritual of the enacted Christian story, focuses the efforts of community toward powerful emotionally attuned experiences that deepen the formation of their participants (Brown & Strawn, 2012). While not exclusive to worship, the particularity of this social activity may warrant particular consideration for its formative ability in such concentrated periods of time.

Mirror neurons provide one neurological basis for the amplification of worship experiences in community. Community, however, might also provide an additional level of causation, or influence, that might shape worshippers. This final level of influence rests upon the most speculative aspects of social neuroscience, but a perspective that bears consideration. One of the key challenges in understanding the relationship between the "mind" and "body" rests with models of Cartesian dualism that often leave a "gap" between conscious free will and biological influence (Murphy & Brown, 2007; Brown & Strawn, 2012). Theorists researching nondualistic views of the mind and brain argue that the "mind" exerts causative influence through a process known as emergence which states that the sum of actions in the mind seem "greater than" their constituent neurological components. Hence the mind, though deeply a part of human creation, displays governing capacities that emerge from the complexity of neurological interaction (Clayton, 2004).

Emergence, as a concept that describes the act of the whole providing causation that cannot be attributed to individual actions, may also describe worship. Some theorists argue that consciousness reflects hybridity between the mind and culture (Donald, 2001; Lunde-Whitler, 2015). People join embodied minds through the practice of worship through a focused activity of narrative, ritual, mimesis through mirror neurons, and affective attunement. The collective activity may result in hybrid connection between the person and the worshipping community (Brown & Strawn, 2012). This collective action of the community may also result in emergent properties (a worshipping consciousness) that directly influence the minds of individual persons.

This model of communal causation respects a different way of understanding how God works through downward causation as well as the quantum level suggested by Markham (Clayton, 2004). Emergent causation, via worshipping practice by the collective community, warrants consideration based on both pneumatological and sacramental perspectives. Practical theologian James Loder (1998) championed the idea of continuity between the human spirit and the Holy Spirit. Drawing upon a metaphor of the Mobius strip (a continuous loop twisted to reveal two interconnected loops), Loder believed there was a persistent, though asymmetrical, relationship between the Holy Spirit and the human spirit. Loder's work, focused upon the life course, tended to focus on individuals, emblematic of his own training in analytic psychology and existentialism. The point of intersection, however, between the Holy Spirit and persons may also occur at a communal level through the emergent causation of the worshipping community.

Thinking of God's activity through worshipping congregation may also make more sense sacramentally. Liturgical scholar Katherine Pickstock (1998) argues that epiclectic movement of the Holy Spirit during communion actually transforms the worshipping community into the body of Christ. Rather than focusing exclusively on the transformation of the elements (bread and cup), the Holy Spirit seeks to "sanctify" the community and, thereby, transform persons as well. Rather than reducing the Holy Spirit's formative power to personal activities, the work of the Holy Spirit "in, through, and under" the causative influence of the worshipping community may reflect a deeper theological framework for understanding the formative power of worship. The result, worshippers participate in the "liturgical construction" of the self (Blevins, 2003). While controversial, this view of the emergent, causative power of the worshipping community upon persons merely accentuates the potential power to form people as they participate in congregational worship.

Regardless of emergent causation, worship represents a powerful model of formation. Undoubtedly the quality of liturgical practice, the nature of the narratives provided, and the attendant focus upon God, or personal taste, will undoubtedly influence the nature of that formation. In addition, as James K. Smith notes, other cultural communal forms may prove even more formative in person's lives, from sports to politics to consumerism. Worship, from a social neuroscientific perspective, may prove more powerful than many Christian educators realize. The combination of practice, narrative, and communal engagement may create the type of social influence rarely replicated in small group bible study or service projects, and unique compared to personal devotional practices. From the smallest synaptic connection, to the collective influence upon the hybrid mind, worship offers the possibility of formation commensurate with the body of Christ, the church.

DISCUSSION QUESTIONS:

1. How does experience shape neurons, why would this be important as we consider educational practices?
2. Describe some of the key stories and practices that define worship in your congregation. Are there clues that identify when worship seems "true" and "virtuous" in that community?

3. Can you describe a time when worship seemed "more" than a group of individual actions?
4. Have you experienced the influence of mirror neurons at a sporting event or worship service? How would you describe it? What difference does it make for us to consider how minds are "connected" to other minds?
5. When you think of the term "body of Christ," what comes to mind? How does our interpretation shape the way we think of both community and personal formation?

REFERENCES

Blevins, D. G. (2002). Healing grace and leadership: Analytic psychology and community. *Wesleyan Theological Journal, 37*(2), 153–171.

Blevins, D. G. (2003). A Wesleyan view of the liturgical construction of the self. *Wesleyan Theological Journal, 38*(2), 7–29.

Blevins, D. G. (2009). Neuroscience, John Wesley, and the Christian life. *Wesleyan Theological Journal, 44*(1), 219–247.

Brown, W. S., & Strawn, B. (2012). *The physical nature of Christian life: Neuroscience, psychology, and the church.* Cambridge, England: Cambridge University Press.

Clayton, P. (2004). *Mind and emergence: From quantum to consciousness.* New York, NY: Oxford University Press.

Cozolino, L. (2013). *The social neuroscience of education: Optimizing attachment and learning in the classroom.* New York, NY: W.W. Norton & Company.

Damasio, A. (2000). *The feeling of what happens: Body and emotion in the making of consciousness.* Orlando, FL: Mariner Books.

Damasio, A. (2010). *Self comes to mind: Constructing the conscious brain.* New York, NY: Pantheon Books.

Donald, M. (2001). *A mind so rare: The evolution of human consciousness.* New York, NY: Norton.

Gladwell, M. (2007). *Blink: The power of thinking without thinking.* New York, NY: Back Bay Books.

Goleman, D. (2007). *Social intelligence: The new science of human relationships.* New York, NY: Random House.

Goleman, D. (2011). *The brain and emotional intelligence: New insights.* North Hampton, MA: More Than Sound.

Hogue, D. A. (2003). *Remembering the future, imagining the past: Story, ritual, and the human brain.* Cleveland, OH: The Pilgrim Press.

Iacoboni, M. (2009). *Mirroring people.* New York, NY: Farrar, Straus and Giroux.

LeDoux, J. (2003). *Synaptic self: How our brains become who we are.* New York, NY: Penguin.

Loder, J. E. (1998). *The logic of the spirit: Human development in theological perspective.* San Francisco, CA: Jossey-Bass.

Lunde-Whitler, J. H. (2015). *The mimetic-poetic imagination: How recent neuroscientific and cognitive psychological research suggests a narratival-developmental approach to identity*. Unpublished manuscript, Religious Education Association Annual Meeting.

MacIntyre, A. (1984). *After virtue* (2nd ed.), Notre Dame, IN: University of Notre Dame.

Maddox, R. (1998, Fall). Reconnecting the means to the end: John Wesley's moral affectional psychology. *Wesleyan Theological Journal, 33*(2), 29–66.

Markham, P. N. (2007). *Rewired: Exploring religious conversion*. Eugene, OR: Wipf & Stock.

Murphy, N., & Brown, W. (2007). *Did my neurons make me do it: Philosophical and neurobiological perspectives on moral responsibility and free will*. Oxford, England: Oxford University Press.

Myerhoff, B. (1978). *Number our days*. New York, NY: E. P. Dutton.

Pickstock, C. (1998). *After writing: On the liturgical consummation of philosophy*. Oxford, England: Blackwell.

Rizzolatti, G., & Sinigaglia, C. (2008). *Mirrors in the brain: How our minds share actions and emotions*. New York, NY: Oxford Press.

Schutt, R. K. (2015). The social brain in a social world. In R. K. Schutt, L. J. Seidman, & M. S. Keshavan (Eds.), *Social neuroscience: Brain, mind, and society* (pp. 231–347). Cambridge, MA: Harvard University Press.

Schutt, R. K., Seidman, L. J., & Keshavan, M. S. (2015). Changing perspective in three disciplines. In R. K. Schutt, L. J. Seidman, & M. S. Keshavan (Eds.), *Social neuroscience: Brain, mind, and society* (pp. 1–28). Cambridge, MA: Harvard University Press.

Siegle, D. L. (2012). *The developing mind* (2nd ed.). New York, NY: Guilford Press.

Smith, J. K. A. (2009). *Desiring the kingdom: Worship, worldview, and cultural formation*. Grand Rapids, MI: Baker Academic.

UMC. (1989). UMH General Thanksgiving I. *United Methodist Hymnal* (pp. 6–12). Nashville, TN: Abingdon Press.

Willis, J. (2010). The current impact on teaching and learning. In D. A. Sousa (Ed.), *Mind, brain, & education: Neuroscience implications for the classroom* (pp. 249–269). Bloomington, IN: Solution Tree Press.

RECOMMENDATION FOR FURTHER READING

Brown, W. S., & Strawn, B. (2012). *The physical nature of Christian life: Neuroscience, psychology, and the Church*. Cambridge, UK: Cambridge University Press.

Cozolina, L. (2013). *The social neuroscience of education: Optimizing attachment and learning in the classroom*. New York, NY: W.W. Norton & Company.

Hogue, D. A. (2003). *Remembering the future, imagining the past: Story, ritual, and the human brain*. Cleveland, OH: The Pilgrim Press.

Schutt, R. K., L. J. Seidman, & M. S. Keshavan (2015). *Social neuroscience: Brain, mind, and society*. Cambridge, MA: Harvard University Press.

CHAPTER 9

MAKING CONNECTIONS

Neurobiology and Developmental Theory

Theresa O'Keefe
Boston College, School of Theology and Ministry

Christian religious educators can draw helpful insights concerning adolescence when we intersect research from the field of neurobiology with developmental psychology frameworks. In particular, recent findings in neurobiology describing changes in the teen brain may help us appreciate the "mechanics" behind claims of developmental psychology. Considered together we gain some insight on the cognitive tasks of adolescents and how we might work effectively to help adolescents mature. Both fields indicate that growth is neither inevitable nor brief, but supported by challenges and supports within context. In this chapter I briefly identify key neurobiological changes happening throughout adolescence. Next, I identify changes possible during adolescence as found in Robert Kegan's framework of development psychology. By way of illustrating the possible changes I look to an adolescent's growing ability to move from identify right and wrong *behaviors* to an ability to recognize the *values* that inform those behaviors. I finish with brief recommendations for Christian educators.

NEUROBIOLOGICAL CHANGES

The human brain matures upward from its base at the brainstem, which controls basic movement like heartbeat and breathing, and eventually finishes with the *prefrontal cortex*, which allows for ideation, historic thinking, long-range planning, and executive functioning. Along the way, all areas of the brain develop, organize, and reorganize until the brain meets full maturity sometime in the third decade of life. One significant change through adolescence is the shift in the decision-making center of the brain. In childhood, decision making is focused in the *amygdala*, which is also the emotional center of the brain. The *amygdala* allows for speedy processing of emotion-laden data such that the child can react quickly to outside stimuli. Decisions made through the *amygdala* are primarily self-serving and instrumental: How can I get what I need/want? How do I avoid what I do not want? It depends on the visible, immediate, and concrete, like objects, emotions, and behaviors, to make decisions. This capacity, although somewhat limited, is essential for the child to function safely with some freedom and reliability in the world. That decision-making capacity has the potential to expand beyond the immediate and concrete when the brain grows and reorganizes through adolescence.

At the onset of puberty there is a burst of growth in the brain (Giedd et al., 1996; Giedd, 1999; Johnson et al., 1999; Lenroot & Giedd, 2006; Spear, 2010). New cells grow in the *prefrontal cortex,* which await assignment and organization over the next decade. There will also be growth of the *corpus callosum,* the network of cells that connect the left and right hemispheres, and further development of the *cerebellum* that seems to assist brain processing overall. All this signals the beginning of a shift in how the brain will function as it moves into adulthood. With time, the new cells organize, new neural pathways grow and strengthen, and unused cells eventually die off, thus enhancing efficiency. These new connections across hemispheres aid access and communication among regions of the brain. The full shift takes years to accomplish. These changes mean that the brain can learn new things, become better able to access information stored in the brain, make connections across regions, and become more reflective and less reactive in the thinking process. Throughout adolescence, the brain's decision-making center moves away from the *amygdala* to the *frontal cortex,* thus allowing for reflection, critical thinking, historical memory, and imagination. In a very real way, the frontal cortex allows the growing person to step back from the immediacy of the *amygdala's* reactions to offer a more reflective and plan-full response to situations.

Adolescence presents a unique window of opportunity for new learning, including new ways of thinking. As the brain becomes newly organized this extensive period is ripe for learning and mastering new cognitive tasks, especially those associated with recognizing history, reasoning, ideas and themes. For our discussion I focus on how adolescence is the time in life to develop a sense of moral reasoning. Such reasoning draws together the capacities of ideation, history, and imagination in order to recognize meaning/intent and long-term cause-and-effect. In a very real way, the brain grows and adapts to meet the challenges presented; it becomes newly "wired" so as to enable the adolescent to be more self-directing in a complex world. Yet these capacities are not brought on board by "flipping a switch," nor is there anything inevitable about this process. Rather, new neural pathways are grown and strengthened through repeated use, so the tasks the adolescent experiences and repeats are those that get developed and strengthened. Neuroscientist Jay Giedd speaks of it as the "lose it or use it" principle ("Interview: Jay Giedd," 2002, para. 13). If opportunities are taken, the young person can move beyond the limits of the immediate and concrete, to recognize and become adept at conceiving ideas and themes, time and space. Sometime in the mid-20s, the process slows and the brain reaches its adult form; the pathways that are used have been strengthened and those that are not have been pruned (Lenroot & Giedd, 2006). As we come to appreciate the growth happening within the adolescent brain, we can take advantage of this valuable window of opportunity. We can invite the adolescent to see and make sense of the world with greater texture, complexity, and meaning.

PSYCHOLOGICAL DEVELOPMENT: SUBJECT-OBJECT THEORY

Brain research offers a physiological rationale for the shifts in meaning-making as described by developmental theorist Robert Kegan. His framework of *constructivist developmental psychology* speaks to how people construct meaning as they mature (Kegan, 1982, 1994). Kegan's research is based on *subject-object theory*, which is a means of identifying cognitive capacity by recognizing what a person is "subject to" and what they can see as an "object" in their lives. That which is *object* is recognized; it can be worked with and manipulated. One has the potential to make decisions about *objects*. On the other hand, *subject* is that which is unseen, given, and unalterable; one simply lives with what is *subject* (Kegan, 1994, p. 32). For example, a person for whom actions are *object* (and values *subject*) believes the actions are good or bad simply because some authority said so, but she has no concept of the acts being morally good or bad. The values, which inform the action, remain invisible *subject*. In such a case she can be expected to behave well if she is monitored, or fears retribution for getting caught, or wishes to please the person giving instruction. On her own she does not know what makes the act right or wrong. On the other hand, when she is able to hold values as *object*, she is able to recognize how they inform actions. Thus she has greater agency in choosing to respond as is fitting to the moment; her actions can reflect the values she holds. She has the capacity for moral agency.

Kegan, building off the earlier work of Piaget, suggests that people develop a capacity to disembed from all manner of things they are *subject* to, such that they can see them with perspective and make decisions concerning them. Kegan names different classes of *subjects* which may move to functioning as *objects* as one comes to see and manipulate them as such. He frames these different classes within five "orders of consciousness." Within his framework, as one moves from one order to the next, the class of elements that were subject in one order, become object in the next (Kegan, 1994). Kegan identifies the first two orders as happening in the course of childhood; movement toward the third is initiated in adolescence and carries into adulthood. The fourth and fifth are possible—but not inevitable—in later adulthood. For the purposes of this chapter, I focus on the transition from Kegan's *second order consciousness* (that of *durable categories*) to *third order* (*cross-categorical meaning making*), which usually transpires as adolescents journey out of late childhood toward adulthood. It includes the shift from *ideas* and *themes* being *subject*, toward the capacity to recognize and work with *ideas* and *themes* as *object*. As such, it is dependent on the development and use of the *frontal cortex* so that the thematic and reflective is possible.

MASTERING THE CONCRETE—SECOND ORDER CONSCIOUSNESS

According to Kegan, the *second order* comes into play as the child is able to recognize the difference between concrete reality and fantasy. Within this order the world becomes one of "durable categories"—the concrete or immediate become observable and manipulatable *object*. Children begin to see the world as learnable and in that way, reliable and consistent. They are also able to see and control their impulses, such that they can respond to parents or teachers demands for good behavior. This child can learn about the world around them; they can learn to follow instructions; they can take control of their actions. This illustration shows the abilities but limits of second order:

> I'll never forget when I snapped at my nephew Max and begged him for just one minute of quietness. He said, "Okay" as he turned around and walked out the door. Then he stopped to inform or ask me again, "Aunt Laurie, one minute is actually sixty seconds. Do you really want one minute or a few

minutes? Do you want me to come back and tell you when the time is up or will you come out and play with us?"

Max at 5 years old is very good at knowing what a minute is, even well enough to instruct his aunt and suggest she may want more time than sixty seconds. However, this episode also indicates the limits of a second order thinker; he cannot infer her intent for an extended period of solitude. According to Kegan, for the second order thinker, concrete and observable items, instructions, and even people, become *objects* which they can make sense of and manipulate. They can see and categorize concrete and immediate things—like Legos and minutes on a clock. However, what they cannot see—what they are still *subject* to—are the *intent* or *value* behind the concrete. What they cannot see are the themes or abstractions which may unite, connect, or order the concrete. For example, Max cannot really imagine the values that inform a request for time away from him. As a young second order knower, Max sees the world from his own perspective and imagines that hers is very much like his own. As Max gets a little older and savvier, he will be able to recognize that Laurie has different interests, but he will still assume that personal interests are what determine behavioral choices (Kegan, 1994, p. 28). Concepts like shared values, which direct or inform actions, remain invisible.

As a means of surviving in the world, a second order thinker is always looking to manipulate the various *objects* of his or her life to meet the point of view that he or she sees as operative. For 5-year-old Max it means waiting quietly for 60 seconds in order to get back to playtime with his aunt. At 15 it is more likely to be more savvy and nuanced: doing what it takes to make mom or dad happy enough to get permission to use the car, or get a new phone. In both cases it is a tit-for-tat sense of cause and effect; "I do what you want so I can get what I want." In that sense, these *objects* (demands, information, people, etc.) become obstacles that the second order knower learns to navigate around—like a simple maze—in order to reach a destination. As children age and move into their teen years, they get better at seeing those *objects* and better at moving around them; they can work longer and more complicated mazes. What in a younger child are necessary skills of self-preservation appear in a young adolescent as selfish and self-centered. The adults around an adolescent recognize such a limited frame of reference as inadequate for the expectations we have of them as they grow. We give the 5-year-old Max the benefit of the doubt that he does not understand his aunt's request and we laugh at the humor of it. We have less patience when Max is 18 and he does not understand and appreciate the intention behind our requests; and we are frustrated with what we think is his thoughtlessness.

LIMITS OF THE CONCRETE—THIRD ORDER

Being able to recognize value or intent is the growing ability of a third order knower. As Kegan writes, they can "subordinate [these concrete *objects*] to the interaction between them" such that they can think abstractly and self-reflectively (Kegan, 1994, p. 29). The ability to conceive of intent, design, or values can evolve in time, and requires that the adolescent depend increasingly on the *frontal cortex*, so that she can see and reflect on ideas or themes and be less dependent on emotions, quick impressions, and immediate benefit. However, this ability is not automatic with age; the ideas and connections have to be learned. Christian Smith, in *Lost in Transition*, a study of 18- to 23-year-olds, offers several illustrations of older adolescents unable to perceive value. Note in the following exchange how a moral framework remains invisible to the interviewee:

I: What is it that you think makes something right or wrong?
R: I'm not sure, I guess probably what it is and stuff like, I don't know.
I: Do you think it has to do with consequences or laws or what?
R: I don't know. I really don't know. (Smith, 2011, p. 36)

I suggest the interviewee does not even know what the interviewer is asking, even when prompted with the idea of "consequences or laws." Similarly, Smith recounts other interviews wherein young adults could not reference moral frameworks. Rather, Smith observed that whether acts were right or wrong were judged by "whether or not anything *functionally improved people's situation*" (Smith, 2011, p. 38, italics in original). Note that this respondent focuses on concrete circumstances and feelings, not on anything thematic, such as values or morals: "Like what will be more fun or what will make my friends have a better time or what will make everyone in the situation … just like happy together, what will be less, least tense situation, what will ah, um, [*pause*] I don't know. Yeah." (Smith, 2011, p. 38).

Also notably, their frame of reference is limited to him and those he knows. The imaginative frame of reference is seldom beyond the immediate, indicating the capacity of second order, in which the concrete is knowable *object* and values remain invisible *subject*. To return to the metaphor named above, as children age to adolescence they can work more complicated mazes and figure out who controls the maze, but they will not wonder why they are in a maze or what informed the design of the maze, or if the maze makes any sense. The maze just is.

BEGINNING TO SEE THE INVISIBLE

Clearly as adolescents move toward adulthood they are expected to take on greater responsibility for themselves and the lives of others. They are expected and depended upon to act independently for their sake and for the sake of others around them. Thus, it is important that they develop the capacity to think and act morally, regardless of whether someone is watching or threatening punishment. They need to develop the capacities of third order knowing. Besides thinking abstractly and self-reflectively, the *third order knower* can subordinate their personal desires and point of view to "loyalty and devotion to a community of people or ideas larger than the self" (Kegan, 1994, p. 32). They can appreciate and learn the requirements of building a relationship—such as trust, responsibility, sympathy, and consideration—without having to receive concrete instructions on how those are met or without serving their own needs over the needs of the relationship. More than simply following specific directions that result in moral behavior, they can *think morally* about behavior. As Kegan writes, they are able to "take out membership in a community of interest greater than one, to subordinate their own welfare to the welfare of the team" (Kegan, 1994, p. 47).

The following quote from one of Smith's interviewees is a good example of someone who has gained that capacity:

> It doesn't matter if you get caught or not. You should have the conscience to say that it's not okay, regardless of whether someone else saw you do it or not. At the end of the day you did it, so I mean just because you can run a red light doesn't mean you should just because there's no cops around. I don't agree with that at all. I don't like that kind of outlook. (Smith, 2011, p. 53)

The reference to "conscience" and the expectation for good behavior regardless of being caught indicates that these things have become *object*, about which this young person can make decisions. She can recognize behaviors within a framework of values and morality.

HELPING YOUTH SEE

Third order cognition exercises rewired the brain, especially its capacity for ideation, reasoning, reflection, long term cause and effect, and connecting different concrete information categorically. Neural changes of adolescence contribute to the capacity to think reasonably; it allows the newly developing brain to ideate and connect concepts thematically. Such growth, for example, makes it possible for the adolescent to recognize that Christian teachings on morality are not simply a list of arbitrary prohibitions (e.g., Thou shalt not murder, lie, steal) but expressing particular values (e.g., the dignity of human life and human community). How might an adolescent make the shift from not seeing these values to seeing them? By mature adults making the thinking more transparent and connected to the concrete realities the adolescent *can* recognize. Let me close with a few suggestions.

- *Make connections transparent.* Starting with examples of the concrete (e.g., Christian practices; actions deemed good or bad), gather them together to illustrate how they contribute to an important ideas—such as Christian virtues or values. Furthermore, refer to the sources that inform the idea, like the Bible or other church teachings. Invite adolescents to practice reading and working with the sources themselves to appreciate how the interpretive process works.
- *Use words.* As adolescents begin to move within the realm of ideas, words become important as never before. Language is the primary tool by which ideas and intentions are communicated and understood. Beyond simply reading or hearing words, it is important that the process be more dialogical or conversational so that the adolescent has an opportunity to train his own voice and his own ideas in the process.
- *Model practices.* The actions of the mature adult community become a means of communicating values upheld and embodied. Rather than appear hypocritical—professing one thing and doing another—the model of the community's action undergirds the instructions expressed. Better yet, invite the adolescent to participate in the practices together with the adults. Invite them to participate in the virtue and values of the community with the wider community.
- *Repeat.* This kind of learning will take more than one instance or conversation. Instead multiple opportunities whereby lessons are restated and reinforced are necessary. Eventually the adolescent will come to see the connections between the actions. With assistance and time, the adolescent can see and make sense more independently and reliably. In this way we teach the young person how to *think and act* as a Christian even when confronted with new situations.

Both the neurologists and Kegan argue that growth comes through a balance of support and challenge—not one or the other alternately, but a mix of both (Lenroot & Giedd, 2009; Kegan, 1994). The adolescent is neither left completely to her own devices, nor is the task completed for her. Rather, she is coached in such a way that the mature considerations around the task are made more transparent. She will need time to see the themes and ideas with reliability.

DISCUSSION QUESTIONS

1. Describe a moment where you observed an adolescent standing on the threshold between different ways of making sense. What did you notice? How do you make sense of it now?
2. Identify a moment when a young person was not been able to see beyond his own perspective. How would you help him recognize or imagine the point of view of another?
3. As you think about values that you wish a young person to learn, how might you articulate the connections between concrete behaviors and the values they support?
4. Recall a situation when an adolescent seemed confused by the nature of a discussion or questions. Recall how she talked about what was happening. What was she able to name and what seemed "invisible" to her but obvious to you? What was object? What remained subject?
5. Consider favorite religious practices (e.g., attendance at worship, sacred reading). What are the values that you find come through those engaging in those practices? Talk about the connection of the practices with the values as you experience them.

REFERENCES

Giedd, J. (2002). *Frontline: Inside the teenage brain.* Retrieved from http://www.pbs.org/wgbh/pages/frontline/shows/teenbrain/interviews/giedd.html

Giedd, J. N., Rumsey, J. M., Castellanos, F. X., Rajapakse, J. C., Kaysen, D., Vaituzis, A., ... Rapoport, J. L. (1996). A quantitative MRI study of the corpus callosum in children and adolescents. *Developmental Brain Research, 91*(2), 274–280.

Giedd, J. N., Blumenthal, J., Jeffries, N. O., Castellanos, F. X., Hong, L., Zijdenbos, A., ... Rapoport, J. L. (2009a). Brain development during childhood and adolescence: A longitudinal MRI study. *Nature Neuroscience, 2*(10), 861–863.

Johnson, S. B., Blum, R. W., & Giedd, J. N (2009b). Adolescent maturity and the brain: The promise and pitfalls of neuroscience research in adolescent health policy. *Journal of Adolescent Health, 45,* 216–221.

Kegan, R. (1982). *The evolving self: Problem and process in human development.* Cambridge, MA: Harvard University Press.
Kegan, R. (1994). *In over our heads: The mental demands of modern life.* Cambridge, MA: Harvard University Press.
Lenroot, R. K., & Giedd, J. N. (2006). Brain development in children and adolescents: Insights from anatomical magnetic resonance imaging. *Neuroscience and Biobehavioral Reviews, 30,* 718–729.
Smith, J., Christoffersen, K., Davidson, H., & Herzog, P. S. (2011). *Lost in transition: The dark side of emerging adulthood.* New York, NY: Oxford University Press.
Spear, L. P. (2010). *The behavioral neuroscience of adolescence.* New York, NY: W. W. Norton.

SUGGESTIONS FOR FURTHER READING

Elkind, D. (1998). *All grown up and no place to go: Teenagers in crisis.* Boston, MA: De Capo Press.
Frontline. (2002). *Frontline: Inside the teenage brain.* Boston, MA: WGBH. Available at: http://www.pbs.org/wgbh/pages/frontline/shows/teenbrain/
Kegan, R. (1994). *In over our heads: The mental demands of modern life.* Cambridge, MA: Harvard University Press.
Mezirow, J. (Ed.). (2000). *Learning as transformation: Critical perspectives on a theory in progress.* San Francisco, CA: Jossey-Bass.
Spear, L. P. (2010). *The behavioral neuroscience of adolescence.* New York, NY: W.W. Norton.

CHAPTER 10

NEO-PIAGETIAN INSIGHTS INTO NEUROLOGICAL DEVELOPMENT OF YOUNG ADULTS WITH IMPLICATIONS FOR UNDERGRADUATE THEOLOGICAL EDUCATION

James Riley Estep, Jr.
Lincoln Christian University, Lincoln, Illinois

John David Trentham
Southern Baptist Theological Seminary, Louisville, Kentucky

Jean Piaget (1896–1980) remains one of the most influential developmental theorists with regard to cognitive development. His theory directly influences Christian education both practically and theoretically on a broad spectrum of topics from learning to curriculum to instruction and even to influencing our understanding of spiritual formation. However, Piaget's successors, the neo-Piagetians, make significant revisions to Piaget's original theory, partially in response to critics of Piaget's theory. Neo-Piagetian theory, or theories, does not restate or refine Piaget's original theory, but offers a more holistic approach to cognitive development, particularly in childhood, primarily through the integration of other developmental perspectives (Pascaul-Leone & Smith, 1969).

Neo-Piagetians rely on certain assumptions, presuppositions, and stage-oriented developmental trajectories that Jean Piaget asserted, but do not necessarily agree with the finer points or conclusions he reached. Also, they do not limit human development to a rigid structuralism, but utilize insights from alternative theories, such as proposed by Lev Vygotsky. Piaget emphasized the "hardware" of cognitive development and theorists like Vygotsky emphasized the "software," whereas the neo-Piagetians utilize insights from both for a more comprehensive or holistic approach to cognitive development.

As the first adult stage, 18–25-year-olds encounter a distinct set of neurological developments (cf. D'Esposito, 1999; Crone et al., 2006). Neo-Piagetian theory asserts that the structural formation of the human brain does not cease in early adolescence, as Piaget concluded. The brain continues to develop into adulthood, which is the advantage of the neo-Piagetian model of cognitive development (Paus et al., 2001; Popp & Portnow, 2001; Mueller et al., 2008; Steinberg, 2014, p. 18). These continued neurological developments create the need for additional stages of cognitive development, beyond Piaget's formal operations, starting at age 12.

Several theorists present stages of cognitive development beyond the formal operations of Piaget. William G. Perry's nine-step model traces the increasing complexity of decision making (Perry, 1970; cf. Kitchener & King, 1981). Kegan's theory of cognitive development in adulthood describes ways of knowing (Popp & Portnow, 2001, pp. 48–53). These models, especially Perry and his successors, provide an accurate phenomenological description of growth and maturity manifested in experience. Neither Perry nor others, however, had the opportunity to link their formulations to findings from the realm of neuroscience.

The study of the neurological structure of the brain between 18–25 years of age exert distinct influence on our understanding of the cognitive development and behaviors of young adults, a time that coincides with students traditionally entering Christian higher education. Hence, one dimension of neo-Piagetian theory includes the integration of contemporary neuroscience, cognitive development theory, and, to some degree, education (Fischer et al., 2007; Worden, Hinton, & Fischer, 2011; Steinberg, 2014), particularly through the work of Harvard University professor, Dr. Kurt W. Fischer.

THE PERRY SCHEME AS A PHENOMENOLOGICAL FRAMEWORK FOR NEUROLOGICAL DEVELOPMENT

William G. Perry, Jr. serves as the foremost neo-Piagetian theoretician in the sphere of educational psychology generally, and college student development particularly. The nature and pattern of growth that Perry described phenomenologically, Fischer observes and describes neurologically. Perry may serve as a theoretical backdrop for Fischer's observations, or Fischer can provide a neurological framework for Perry's phenomenological description.

William Perry's landmark publication, *Forms of Intellectual and Ethical Development in the College Years: A Scheme*, represents the standard for identifying and explaining epistemological growth in college students through connections between the learner, subject matter, and process of knowing (Moore, 2002, p. 18). One of the core underlying principles evident in the scheme insists that there are limits to formal logic and reason (Butman & Moore, 1997). Perry concludes that beliefs remain inherently contextual and relativistic and require faith commitments. Growth, for Perry, entails the liberalization of a student's perspective, from epistemological absolutism to contextual relativism. Successful cognitive growth entails an increasing, convictional commitment to one's own values and assumptions (formed on the basis of a critical and reflective criteria of assessment), while remaining open to revision of one's worldview through continual testing and discernment in light of alternate, potentially valid truth claims.

Perry as Neo-Piagetian

The Perry Scheme offers a cognitive-structural theory, formulated according to the Piagetian developmental paradigm. While drawing directly from Piagetian premises, Perry extended and deepened the work of Piaget. Piaget's model describes the development of structures and processes that characterize formal logical thinking. Perry's scheme provides a closer examination of how students think about knowledge and authority by mapping the epistemological perspectives that profoundly affect the processes of learning and personal identity development (Kurfiss, 1994, p. 166). Perry provides a "soft" cognitive-structural model which does not fit strictly into Piaget's ("hard") theoretical mold (Moore, 2002, p. 31). Perry identifies his study as reflecting the process articulated by Piaget (assimilation and accommodation) but extends the period of formal operations (what Piaget anticipates as "might be") in a unique exploration (Perry, 1970, p. 228).

The Scheme

As a cognitive-structural model, the Perry Scheme describes the way in which people think and view the world (i.e., their *forms* of epistemology) rather than assessing the specific content of people's thoughts (Butman & Moore, 1997). Perry characterizes the scheme as a "Pilgrim's Progress" of development, thus indicating the progressive, intentional nature of a student's growth, propelled by unsettling experiences which necessitate active commitments in one's approach to identity, knowledge, learning, and values (Perry, 1981). The scheme may be summarized according to four fluid periods: Dualism, Multiplicity, Contextual Relativism, and Commitment within Relativism (Moore, 2002, pp. 20–22).

"Dualism" refers to a student's conception that there is a reality composed of stark dualities, such as between truth and falsehood, right and wrong, good and bad, and so on. These students live in a world of unquestioned, objective absolutes. Their concept of knowledge and learning entails "having the right answer." Dualism includes the assumption that the ultimate source of authority is located outside the self (Parks, 1991). Dualistic thinkers tend to be conformists with identities defined by what others think of them. This perspective shifts drastically in the following period.

As a student progresses to the period of Multiplicity, he or she recognizes the reality of diversity and uncertainty, and the possibility that there may be more than one solution or point of view (Butman & Moore, 1997). Perry suggests a better term for Multiplicity may be "personalism," to describe this simplistic, subjective, structure of cognition. A student recognizes (legitimizes) uncertainty, but lacks the capacity to comparatively analyze different perspectives (Perry, 1981). An epistemically irresponsible attitude emerges that glorifies personal opinion (Perry, 1970). Multiplicity is thus still a form of dualistic thinking, as the student considers knowledge and truth in terms of absolute propositions and values—substantiated completely according to one's own arbitrary inclinations.

The transition from multiplicity to contextual relativism represents the most significant and consequential moment of epistemic maturation in the Perry Scheme, with the possibility of further growth. Students undergo a fundamental worldview reorientation, from viewing knowledge and values in dualistic terms (with an ever-increasing number of exceptions to the rule), to recognizing that knowledge and values remain essentially contextual and relativistic. Perry identifies this reorientation by the discovery of metathought, or the capacity to examine thought, including one's own (Perry, 1981). Rather than "proving" certain propositions absolutely true, theories offer interpretations of reality that can be compared to other interpretations. Contextual Relativism entails a student's realization that reasonable people will reasonably disagree an "inescapable uncertainty," as the only viable epistemological lens by which to evaluate theories and truth claims. Since essentially a cerebral (i.e., not affective) position, Perry calls Contextual Relativism the "space of meaninglessness between received belief and creative faith" (p. 92).

A full awareness, and a responsible acceptance, of the implications of internally based meaning emerges into the final period of development, Commitment within Relativism. Young adults acknowledge the necessity of making responsible (intellectual and moral) judgments within a relativistic world (Moore, 2002). Perry says that positive growth through this final stage requires faith commitments, which at first often involves "arbitrary faith," or "the willing suspension of disbelief" (Perry, 1981, p. 92). The shift from "knowledge and reason" to "faith and commitment" represents an orientation toward the inescapable, ethical dimensions of epistemology.

Perry suggests that the only means of escaping the loneliness that comes with personal commitments in a relativistic world resides in community. (Perry, 1981). Perry believes epistemological vulnerability represents key element of the higher ranges of human development. One reflexively recognizes that intellectually honest and personally authentic commitments must be held with both simultaneous conviction and tenuousness.

From Perry to Fischer

With Perry's phenomenological description as a backdrop, we will (1) itemize the neurological development that occurs in the human brain between ages 18 and 25, (2) identify the impact of these developments on cognition, and (3) assess the implications for Christian higher education in and beyond the classroom. Integrating the neuroscience insights of Kurt W. Fischer with Perry's descriptive analysis leads to implications for those serving in undergraduate Christian education.

NEUROLOGICAL DEVELOPMENT IN EARLY ADULTS

What is actually happening in the brains of 18–25-year-olds? Without oversimplifying, three incredible developments occur. An increase of "white matter" in the prefrontal cortex, a decrease, or "pruning," of neurological connections, and an increase in the cross hemisphere connections in the corpus callosum.

Increase in White Matter

The developments in the prefrontal or orbitofrontal cortex involve executive decision making, especially in regard to moral decisions (Wolfe,

2010). Simpson (n.d., p. 14) calls this area of brain the "executive suite" because it is the control center for so much of what occurs. Between ages 18 and 20, the brain experiences a spurt of development in the frontal-temporal lobe (Fischer & Rose, 1996). Gray matter is the brain's cortex containing nerve cell bodies vs. white matter—which is the part of the brain containing myelinated nerve fibers, which insulates the nerve to insure that the impulses sent by axons are not misdirected and speeding up the transmission of signals (Wolfe, 2010).

This difference separates young adults from older adults as well as from adolescents. The frontal lobes of young adults (ages 18–25) do not function like those of older adults (ages 60–80) in terms of memory. D'Esposito observes, "We can show that the frontal lobes function differently in the two groups during a memory task. Differences in brain activity in the younger and older subjects were limited to one region of the frontal lobes. In this region, older individuals used more of the brain even though they were as accurate as the young during performance of the task" (McBroom, 2000).

Decrease in Neurological Connections

Lenroot and Giedd (2007) note between ages 18–22, male and female brains decline in frontal and parietal gray matter, plateau in temporal grey matter, and experience a marked increase in white matter. Wolfe observes:

> Around age 11, a massive pruning of these connections [in the frontal lobes] begins *and continues into early adulthood*. Although people may assume more synapses would be beneficial, the brain actually consolidates learning by pruning away excess connections and ensuring that the most useful synapses are maintained. This, in turn, allows the brain to operate more efficiently. (Wolfe, 2010, p. 84, emphasis added; cf. also Swentosky, 2008, pp. 9–10)

The improved efficiency of the brain facilitates more skillful and nuanced cognitive faculties in young adults, consistent with Perry's conception of epistemological maturity as being identified by the progression from passive dualism to contextual relativism, and ultimately to authentic commitment (Knight & Sutton, 2004).

Increase in Cross-Hemisphere Connections

Numerous studies indicate the growth of the corpus callosum, and corresponding human cognition, up to the middle 20s (Pujol et al., 1993). However, researchers note that, while the growth ultimately declines with increasing age, the breadth of the size remains statistically significant well into middle age regardless of gender (Pujol et al., 1993). The corpus callosum defines the nerve bundles that connect the two cerebral hemispheres of the brain and facilitate communication between them. The ensuing increase of mental activity in this area of the brain probably reflects "the highest order-latest maturing neural work of the brain" (Pujol et al., 1993, p. 74).

Similarly, Rea Simpson concludes that developmental shifts in young adults represent significant changes in the brain associated with the executive function: "calibration of risk and reward, problem-solving, prioritizing, thinking ahead, self-evaluation, long-term planning, and regulation of emotion" (Simpson, 2009, p. 13). Young adult brains take on the same appearance and process of older adults during the mid-20s.

KURT W. FISCHER'S NEUROEDUCATION THEORY

The actual structural changes within the neurological system of the brain between ages 18–25 should be considered in conversation with the holistic development of the older adolescence. Fischer's (1985, 2003) basic premise remains simple: Neurological developments in early young adulthood yield increased cognitive capacity (Potential) which in turn allows for greater emotional regulation and eventually social and behavior adaptation (Figure 10.1). Likewise, Perry's conception of epistemological maturation entails the acquisition of knowledge as personal and contextual, ethical and conviction based, and the maintenance of knowledge as critically reflective and collaborative within community and society.

Fischer's work on neuroeducation offers *at least* three basic insights that have the benefit of the study of neuroscience in late adolescence. First, persons entering young Adulthood navigate stages that do not resemble Piaget's terminal formal operations. Second, cognitive development includes "spurts" rather than a smooth development associated with the traditional Piagetian model. Third, cognitive development appears more uneven that the assumed even, universal, cognitive development proposed by Piaget.

Fischer's Adult Stages

Whereas Piaget's theory of development ends with the entry into Formal Operation around age 12 (Piaget, 1972), Fischer proposes that cognitive development occurs in three distinct tiers with each tier comprised of four distinct stages, which would be twelve stages of cognitive develop-

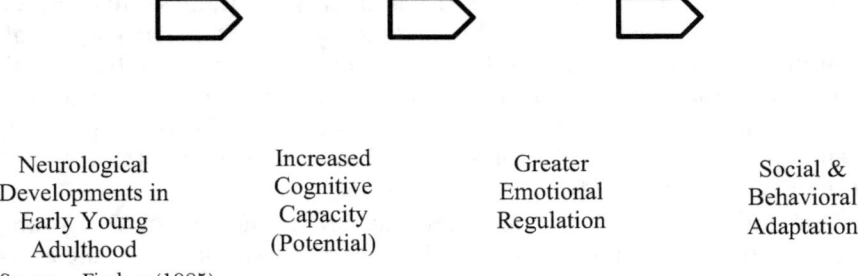

Neurological Developments in Early Young Adulthood → Increased Cognitive Capacity (Potential) → Greater Emotional Regulation → Social & Behavioral Adaptation

Source: Fischer (1985).

Figure 10.1. Basic premise.

ment. However, the tiers share two transitional stages, hence his theory actually consists of ten stages of cognitive development, the latter stages being those of adulthood (Fischer, 1985). Fischer theorizes at least four stages of adult cognitive development, requiring a cumulative ability to consider increasingly complex abstractions. Fischer's more detailed and dynamic neurological research, which Piaget could not conduct, finetuned the original theory's conclusions about cognitive development in adulthood.

Humans develop cognitively progressively from single abstractions to linking the abstractions, to networking multiple linked abstractions culminating in what Fischer described as complete abstraction (Fischer, 1980; Fischer & Rose, 1998; Schenck, n.d.; Commons, Richards, & Armon, 1984). Perry's identification of metathought seems to connect as the engine of epistemological progress. With this ability to think critically, Perry (1981) says,

> Theories become, not "truth," but metaphors or "models," approximating the order of observed data or experience. Comparison, involving systems of logic, assumptions, and inferences, all relative to context, will show some interpretations to be "better," others "worse," many worthless. Yet even after extensive analysis there will remain areas of great concern in which reasonable people will reasonably disagree. (p. 88)

Metathought, "the ability to examine thought, including one's own," provides the necessary competency for epistemic maturation in a relativistic world.

Spurts of Cognitive Development

Fischer detected periods of increased cognitive development, called spurts, which marked increase in the brain's ability to do multiple abstractions from 18–25, followed by a slight decline. Whereas traditional Piagetian theory advocated a smooth, steady, progressive cognitive development, Fischer's model affirms spurts a range in development from optimal, rapid, spurts to functional or normal but gradual development. Heightened cognitive development entails an "optimal level of high support" or holding environment for the student (Popp & Portnow, 2001५; Scheck, n.d., slide 12). While Perry did not describe growth in spurts, he did anticipate this reality through dynamic transitions rather than static stages (Perry, 1981). Also, Perry believed community support to be a prerequisite for promoting maturity (Perry, 1978).

Periods of optimal cognitive development in early adulthood coincide with growth cycles in the brain (Fischer, Yan, & Stewart, 2003). Spurts occur when developmental skills are "leaping forward at a fast pace, especially under conditions that support optional performance" (Fischer, 2008, p. 129). According to Fischer, one of the periods of increased performance begins at age 18, slowing at age 21, increasing again at age 23 (Fischer, Yan, & Stewart, 2003, p. 494; Fischer & Rose, 1996, p. 269).

Uneven Development

Piaget favored uniform, cognitive abilities through developmental stages. Fischer advocates nonuniform, differing cognitive domains may develop at different rates due to changes in the neurological developments in the brain. For this reason, Fisher preferred to describe his model as a web of development rather than a ladder (Battro, Fischer, & Léna, 2010,). Fischer asserts adolescents and young adults develop in the areas of mathematics, self-in relationship, and reflective judgments. One experiences discontinuities in the three skills, yet remains sustained an interactive web of developing cognitive potential (Fischer, 2008). The web metaphor may be coordinated seamlessly with Perry's conception of development and maturity as a Pilgrim's Progress—a whole-life experiential affair that entails dynamic seasons of growth as well as seasons of stagnation and deflection.

Neuroscience and the 18–25-Year-Old Student

Fischer and other theorists note: "People develop differently in separate domains on the basis of their experiences and interests, and their development continues far into adulthood in domains on which they focus" (Fischer, Stein, & Heikkinen, 2009, p. 596). These cognitive developments can differ slightly culture to culture (Morra, 2007). Additionally, advances in moral reasoning may occur during growth cycles/spurts (Fischer & Rose, 1996), a level of moral judgment beyond Piaget's final stage (Commons, Richards, & Armon, 1984). Cognitive developments during this period, from problem solving, critical thinking, deciding, organizing, and strategizing, demonstrate new cognitive capabilities (Wolfe, 2010; Simpson, n.d.).

Vartanian and Mandel (2012) explain, while early adults can make good decisions, their difficulty applying these decisions rest primarily due to neurological developments between the prefrontal cortex's interaction with the limbic system. Young adults do seem to handle emotional intensity better than youth, making "heat of the moment" decisions while considering future consequences (Simpson, 2009, p. 13). Young adults do possess a more accurate perception of self, improved emotional and impulse control, which leads to better decision making, a deeper sense of empathy, and the ability to select behaviors appropriate to a given circumstance (Wolfe, 2010; Simpson, n.d.; Steinberg, 2014).

Implications of Neurological Developments

Based on Fischer's neurological research findings and Perry's phenomenological descriptive analysis, one must clearly recognize the significance of the college and young adult years as paradigmatically influential and formative within God's designed pattern for lifespan development. These years represent the "critical turning point" (see Tanner, 2006) of the movement from dependence to personal autonomy (sociology), from uncritical passivity to reflective judgment (educational psychology). Likewise, findings in neurological research above dovetails with sociological and theoretical descriptions in support of this turning point and telltale progression. Christian young adults in this phase of the lifespan may indeed move from childhood to adulthood for the sake of the gospel, as those "who have their powers of discernment trained by constant practice" (Hebrews 5:14). When we carefully and wisely appropriate findings in neurological research, we honor the dignity, design, and potential of learners as image bearers of God to a greater extent than we otherwise would have before that research was available.

Theological Implications

Neurological maturation processes among young adults reflect and are commensurate with the biblical ideal of sanctification for every believer: ever-increasing transformation unto Christlike wisdom, pursued within authentic community. The premise for positive development in Scripture is defined by the doctrine of sanctification. For believers, sanctification provides the means of obtaining and growing in genuine knowledge—a knowledge of God that is intrinsically personal, relational, and active (Packer, 1993, p. 20). Believers' personal, relational knowledge of God entails a transformative, progressive, pilgrimage of development as Christians are conformed to the *imago Christi* (Rom 8:29). The Bible thus encourages believers to "put on Christ" (Rom 13:14) as they are "transformed into the same image" (2 Cor 3:18).

The biblical expectation of transformative conformity to Christ's image highlights the truth that development is not static, but dynamic (Hoekema, 1994). Believers are *being renewed* (Col 3:10) in the image of God as they await ultimate redemption. The most evidential mark of epistemic virtuosity is growth toward Christlikeness, coupled with a determined commitment to the "upward call" (Phil 3:13–14). Sanctification is never fully obtained (Phil 3:12), but actively pursued (1 Tim 6:11) through obedience and wisdom. Wisdom relates primarily to daily "living skillfully" within God's embedded structure (Estes, 1997, p. 26). For Christians the ultimate aim and purpose of wisdom is the person of Christ, "the power of God and the wisdom of God" (1 Cor 1:24).

Developmental Implications

Sociologists who study young adults argue whether the overarching goal during this phase of life is to "stand on one's own two feet," such as to become independent or self-sufficient (Smith, 2009, p. 150) or "recentering" from parental regulation to self-regulation (Tanner, 2006, p. 27). Neuroscience, sociology, and also Christian worldview, agree that social-environmental influences remain pervasive and consequential in the lives of all people. Human development occurs according to one's experience and cultural context(s) established in the relational aspect of God's *imago Dei*; a concept that defines the uniqueness of personhood and also dependency on relationships. People, imbued with God's relational image, develop according to environmental factors consistent with the social norms of their various communities and contexts. Guided by the indwelling Spirit, Christians develop according to their identification with the "community of faith" (Col 3:9–17) as well as their familial and societal contexts.

Pedagogical Implications

Following the example of Jesus, education that is distinctively Christian must have as its chief end the continual progression of individuals toward maturity in Christ, such that God is glorified and the diverse communities of faith are strengthened in its calling (Dockery, 2007; Estep, 2008). Confessionally oriented Christian teachers facilitate intellectual and spiritual transformation, emphasizing development rather than "accumulation." Dissatisfied with merely a student's grasp of propositions (Bain, 2004), true learning involves a sustained influence on one's actions and feelings, leading to holistic maturation and wisdom. Teachers make every effort to create a learning environment conducive to growth (Bain, 2004). Brookfield argues for classroom *dialogue* that entails an "unpredictable creation of meanings through collaborative inquiry" (Brookfield, 2006, p. 129). Providing a diversity of perspectives in a respectful but challenging community of learners influences the tenor and content of developmental growth.

Discipleship Implications

Finally, young adult neurological maturation supports a paradigm for redemptive development—such as discipleship—rooted in discernment and biblical wisdom, and applied through the ethic of missional Christian love. Redemptive development entails knowledge, understanding, and wisdom; rooted in Christ, propelled by hope, and pursued with missional integrity for the glory of God. The redemption secured by Christ's sacrifice and resurrection is the premise on which the whole Christian concept of life, purpose, and hope rests. Yount (2010) describes the resulting trend of redemptive development as an epistemological progression unto wisdom. This new orientation fundamentally redefines the nature and purpose of development to center on glorifying God through active obedience, proclaiming and extending the same grace one has received. Authentic Christian discipleship is thus holistic and developmental in scope. Holistic human development is characterized as cognitive, affective, and behavioral. The intellectual assent to the truth of the gospel is intertwined with one's affections and convictions, which bears forth practical applications of faith. Holistic development is thus critical to maturity. A mature Christian knows the gospel, resonates with the gospel, and applies the gospel—thus gaining wisdom. Numerous passages could be cited to exhibit the biblical principle of holistic growth, but none is more prominent than the greatest commandment: "Hear, O Israel: The LORD our God, the LORD is one. You shall love the LORD your God with all your heart and with all your soul and with all your might" (Deut 6:4–5).

DISCUSSION QUESTIONS

1. In what ways were Perry and Fischer describing the same episode in adult development?
2. When have you experienced or observed Fischer's form of development, for example, spurts?
3. How might the implications identified in this chapter materialize in your ministry or work?
4. What influenced you most from this chapter in regard to your work with young adults still developing cognitively?

REFERENCES

Bain, K. (2004). *What the best college teachers do.* Cambridge, MA: Harvard University Press.
Battro, A. M., Fischer, K. W., & Léna, P. J. (2010). *The educated brain: Essays in neuroeducation.* Cambridge, England: Cambridge University Press.
Brookfield, S. D. (2006). *The skillful teacher: On technique, trust, and responsiveness in the classroom.* San Francisco, CA: Jossey-Bass.
Butman, R. E., & Moore, D. R. (1998). The power of Perry and Belenky. In J. C. Wilhoit & J. M. Dettoni (Eds.), *Nurture that is Christian: Developmental perspectives on Christian education* (pp. 105–122). Grand Rapids, MI: Baker Books.
Commons, M. L., Richards, F., & Armon, C. (1984). *Beyond formal operations: Late adolescent and adult cognitive development.* New York, NY: Praeger.
Crone, E. A., Wendelken, C., Donohue, S., van Leijenhorst, L., & Bunge, S. A. (2006). Neurocognitive development of the ability to manipulate information in working memory. *PNAS, 103*(24), 9315–9320.
D'Esposito, M. (1999). Cognitive aging: New answers to old questions. *Current Biology, 9*(24), 939–941.
Dockery, D. (2007). *Renewing minds: Serving church and society through Christian higher education.* Nashville, TN: B&H.
Estep, J. R., Jr. (2008). Toward a theologically informed approach to education. In J. R. Estep Jr., M. J. Anthony, & G. R. Allison (Eds.), *A theology for Christian education* (pp. 264–295). Nashville, TN: B&H.
Estes, D. J. (1997). *Hear my son: Teaching and learning in Proverbs 1-9.* Downers Grove, IL: InterVarsity Press.
Fischer, K. W. (1980). A theory of cognitive development: The control and construction of hierarchies of skills. *Psychological Review, 87*(6), 477–531.
Fischer, K. W. (1985). Stages and individual differences in cognitive development. *Annual Review of Psychology, 36*, 613–648.
Fischer, K. W. (2008). Dynamic cycles of cognitive and brain development: Measuring growth in mind, brain, and education. In A. M. Battro, K. W. Fischer, & P. Léna (Eds.), *The educated brain* (pp. 127–150). Cambridge, England: Cambridge University Press.
Fischer, K. W., & Rose, S. P. (1996). Dynamic growth cycles of brain and cognitive development. In R. W. Thatcher, G. R. Lyon, J. Rumsey, & N. Krasnegor (Eds.), *Developmental neuroimaging: Mapping the development of brain and behavior* (pp. 263–279). San Diego, CA: Academic Press.
Fischer, K. W., & Rose, S. P. (1998). Growth cycles of brain and mind. *Educational Leadership, 56*, 56–60.
Fischer, K. W., Yan, Z., & Stewart, J. B. (2003). Adult cognitive development: Dynamics in the developmental web. In J. Valsiner & K. Connolly (Eds.), *Handbook of developmental psychology* (pp. 491–516). Thousand Oaks, CA: SAGE.
Fischer, K. W., Daniel, D. B., Immordino-Yang, M. H., Stern, E., Battro, A., & Koizumi, H. (Eds.). (2007). Why mind, brain, and education? Why now? *Journal Compilation, 1*(1), 1–2.

Fischer, K. W., Stein, Z., & Heikkinen, K. (2009). Narrow assessments misrepresent development and misguide policy. *American Psychologist*, *64*(7), 595–600.

Hoekema, A. A. (1994). *Created in God's image*. Grand Rapids, MI: William B. Eerdmans.

Kitchener, K. S., & King, P. M. (1981). Reflective judgment: Concepts of justification and their relationship to age and education. *Journal of Applied Developmental Psychology*, *2*, 89–116.

Knight, C. C., & Sutton, R. E. (2004). Neo-piagetian theory and research: Enhancing pedagogical practice for educators of adults. *London Review of Education*, *2*(1), 47–60.

Kurfiss, J. (1994). Intellectual, psychosocial, and moral development in college: Four major theories. In K. A. Feldman & M. B. Paulsen (Eds.), *Teaching and learning in the college classroom* (pp. 165–191). Needham Heights, MA: Simon & Schuster.

Lenroot, R. K., & Giedd, J. N. (2007). The structural development of the human brain as measured longitudinally with magnetic resonance imaging. In D. Coch, K. W. Fischer, & G. Dawson (Eds.), *Human behavior, learning, and the developing brain: Typical development* (pp. 50–73). New York, NY: Guilford Press.

McBroom, P. (2000). *Campus unveils powerful brain scanner devoted solely to research*. Retrieved from http://www.berkeley.edu/news/berkeleyan/2000/11/29/mripat.html

Moore, W. S. (2002). Understanding learning in a postmodern world: Reconsidering the Perry scheme of ethical and intellectual development. In B. K. Hofer & P. R. Pintrich (Eds.), *Personal epistemology: The psychology of beliefs about knowledge and knowing* (pp. 17–36). Mahwah, NJ: Erlbaum.

Morra, S., Gobbo, C., Marini, Z., & Sheese, R. (2007). *Cognitive development: Neo-piagetian perspectives*. Florence, KY: Psychology Press.

Mueller, V., Brehmer, Y., von Oertzen, T., Li, S., & Lindenberger, U. (2008). Electrophysiological correlates of selective attention: A lifespan comparison. *BMC Neuroscience*, *9*(18), 1–21.

Packer, J. I. (1993). *Knowing God* (20th ann. ed.). Downers Grove, IL: InterVarsity Press.

Pascaul-Leone, J., & Smith, J. (1969). The encoding and decoding of symbols by children: A new experimental paradigm and a neo-piagetian model. *Journal of Experimental Child Psychology*, *8*(2), 328–355.

Parks, S. (1991). *The critical years: Young adults & the search for meaning, faith, and commitment*. San Francisco, CA: HarperCollins.

Paus, T., Collins, L., Evans, A. C., Leonard, G., Pike, B. G., & Zijdenbos, A. P. (2001). Maturation of white matter in the human brain: A review of magnetic resonance studies. *Brain Research Bulletin*, *54*(3), 255–266.

Perry, W. G., Jr. (1970). *Forms of intellectual and ethical development in the college years: A scheme*. San Francisco, CA: Jossey-Bass.

Perry, W. G., Jr. (1978). Sharing in the costs of growth. In C. A. Parker (Ed.), *Encouraging the development of college students* (pp. 267–273). Minneapolis, MN: University of Minnesota Press.

Perry, W. G., Jr. (1981). Cognitive and ethical growth: The making of meaning. In A. W. Chickering (Ed.), *The modern American college: Responding to the new realities of diverse students and a changing society* (pp. 76–116). San Francisco, CA: Jossey-Bass.

Piaget, J. (1972). Intellectual evolution from adolescence to adulthood. *Human Development, 15*(1), 1–12.

Popp, N., & Portnow, K. (2001, August). Our developmental perspective on adulthood. *NCSALL Reports #19*, 43–69.

Pujol, J., Vendrell, P., Junqué, C., Martí-Vilalta, J. L., & Capdevila, A. (1993). When does human brain development end? Evidence of corpus callosum growth up to adulthood. *Annals of Neurology, 34*, 71–75.

Schenck, J. (n.d.) Discoveries in the adolescent brain: Implications for teaching and learning. Retrieved from knowa@directairnet.com

Simpson, R. (2009). *The role of public service in young adult development: Highlights from recent research.* Unpublished paper, MIT Public Service Center.

Simpson, R. (n.d.) *Young adult development: What the research tells us.* Unpublished presentation, MIT, Parenting Education & Research.

Smith, C. (2009). *Moral believing animals.* New York, NY: Oxford University Press.

Steinberg, L. (2014). *Age of opportunity: Lessons from the new science of adolescence.* New York, NY: Houghton Mifflin Harcourt.

Swentosky, A. J. (2008). *A neo-piagetian approach to social cognitive development* (Master's thesis). University of Pittsburgh, PA.

Tanner, J. L. (2006). Recentering during emerging adulthood: A critical turning point in life span human development. In J. Jensen Arnett & J. L. Tanner (Eds.), *Emerging adults in America: Coming of age in the 21st century* (pp. 3–20). Washington, DC: American Psychological Association.

Vartanian, O., & Mandel, D. R. (2012). Neural basis of judgment and decision making. In M. K. Dhami, A. Schlottmann, & M. R. Waldmann (Eds.), *Judgment and decision making as a skill: Learning, development, and evolution* (pp. 29–52). New York, NY: Cambridge University Press.

Wolfe, P. (2010). *Brain matters: Translating research into classroom practice* (2nd ed.). Alexandria, VA: ASCD.

Worden, J. M., Hinton, C., & Fischer, K. W. (2011). What does the brain have to do with learning? *Kappan, 92*(8), 8–13.

Yount, W. R. (2010). *Created to learn: A Christian teacher's introduction to educational psychology* (2nd ed.). Nashville, TN: B&H Academic.

SUGGESTIONS FOR FURTHER READING

Fischer, K. W., (2003). *Handbook of developmental psychology.* Thousand Oaks, CA: SAGE.

Morra, S., Gobbo, C., Marini, Z., & Sheese, R. (2007). *Cognitive development: Neo-Piagetian perspectives.* Florence, KY: Psychology Press.

Perry, W. G. (1970). *Forms of intellectual and ethical development in the college years: A scheme.* San Francisco, CA: Jossey-Bass.

CHAPTER 11

CHANGING BEHAVIOR AND RENEWING THE BRAIN

Neuroscience and Spirituality

Mark A. Maddix
Northwest Nazarene University

Glena Andrews
George Fox University

The field of neuroscience and religion continues to explode as researchers seek to understand religious experiences in the brain. Studies in religious experience, called *neurotheology*, attempt to draw conclusions about the truth of these religious experiences from the study of biological brain events. On the surface it may seem these recent developments could cause some Christians concern about providing scientific proof for "religious experience." However, truthfully, science could provide a basis for understanding how practices such as meditation and prayer impact the brain.

The question of religious experience and brain activity provides a fascinating field of study. Studies show that whether drug induced, seizure-related, magnetically stimulated, or born of normal brain processes, religious experiences remain clearly tied to physical brains. Some neuroscientists attempt to "locate" religious experience in one particular brain module or wiring system but later realized that different religious experiences impact different parts of the brain (Newberg & Waldman, 2010). However, religion cannot be reduced to a primary form of cognitive activity, such as language. Religion exists more like a cultural and social phenomenon that includes a variety of individual group experiences, events, and activities (Jeeves & Brown, 2009, p. 99).

As noted in a previous chapter, studying the impact of meditation on brain activity provides an area of exploration in neuroscience and religious experiences. Andrew Newberg and Mark Waldman, at the University of Pennsylvania and the Center for Spirituality and the Mind, conducted one of the most extensive studies of neuroscience and religious experience. For several years they studied how different concepts of God affects the human mind (Newberg & Waldman, 2010). They evaluated brain scans of Franciscan nuns as they immersed themselves in the presence of God, and charted the neurological changes as Buddhist practitioners contemplated the universe. They studied the brains of Pentecostal practitioners who invited the Holy Spirit to speak to them in tongues, and viewed how the brains of atheists react, or do not react, when they meditate on a concrete image of God (Newberg & Waldman, 2011). Newberg and others are mapping the neurological changes caused by spiritual and religious practices. Their result has led to the following conclusions:

- Current research in the field of neuroscience and spirituality shows that meditation enhances the neural functioning of the brain in ways that improve physical and emotional health (Newberg & Waldman, 2010).
- Long-term contemplative practices strengthen a specific neurological circuit that generates peacefulness, social awareness, and compassion for others (Newberg & Waldman, 2010).
- Some neuroscientists claim that spiritual formation practices improve memory and can slow down neurological damage caused by old age (Newberg & Waldman, 2010).
- Meditation and contemplation of God can change the structures of the brain that control moods, give rise to the conscious notions of self, and shape the sensory perceptions of the world (Newberg & Waldman, 2010).

Newberg and Waldman's studies show that religious and spiritual contemplation changes the brain in profound ways by changing the neural circuitry resulting in enhanced cognitive health. It shows that long-term contemplative practices strengthen a specific neurological circuit that generates peacefulness, social awareness, and compassion for others. Also, Sharon Begley (2007) believes that the human brain has the capacity to rearrange itself and be changed through meditation. *Neuroplasticity* is the mechanism that allows these changes to occur in the brain. Religious experience is cognitively structured where thoughts and beliefs play a central role. Neuroscience research shows that religious experiences are associated with patterns of brain activity but no specific brain area mediates religious experience.

Given the substantial research on the science of religion, this chapter explores the physiological changes of college students engaged in regular spiritual practices. Students were asked to engage in intentional spiritual formational practices such as prayer, meditation, scripture reading, and contemplation to see if these practices impacted their physiological activity including brain wave, heart rate, skin response (sweat), and reaction time changes. A variety of neuropsychological measures such as the Beck Depression Inventory, State/Trait Anxiety Scale, heart beats per minute, Galvanic Skin response, and Electrophysiological Encephalography measures were gathered at the beginning and end of the course to measure physiological activity. The purpose was to discover that as students developed more mature formation of spiritual disciplines, their physiological measures changed as a result of spiritual formation practices.

CASE STUDY HYPOTHESIS

Students were introduced to *inward* or contemplative practices for the study. They were to engage in an inward practice 20 minutes per day, 7 days a week, for 4 weeks. The practices included: Contemplative prayer, centering prayer, silence and solitude, journaling, meditation, visualization through icons, images, and music, scripture meditation, and *Lectio Divina* (sacred reading) (Calhoun, 2005; Jones, 2005; Leclerc & Maddix, 2011). The focus on contemplative spirituality has often been linked with the mystical dimension of faith to see the inner harmony of all things and denies any dualistic notion of life that opposes spirit and matter, the divine and human. The contemplatives have searched for God in solitude (hermetic tradition) and in communal settings (monastic tradition).

Given the unique setting of a cohort experience and using research conducted on the effects of meditation on the brain, the researchers hypothesized that changes could occur in the physiological and psychological measures of the college students but were not certain that the varied spiritual disciplines, including meditation, done for only 20 minutes each day could result in the same change in brain waves. The independent variable was being in a course where students participated in spiritual disciplines 20 minutes each day for a 4-week period. The dependent variables included pre- and postmeasurements of 12 brainwave sites (mean power), galvanic skin response (microseman) and heart rate (beats per minute), pre- and postmeasurements of anxiety and depression.

Methods

Participants

Students were recruited from a traditional undergraduate university population. Recruiting announcements were emailed to the undergraduate student body at the time of scheduling for the next semesters. The course description was included along with the format for the course structure. The experimental group consisted of male and female undergraduate students ($n = 38$) enrolled in a 7-week course on the history of spiritual practices and neuroscience offered during the first 7 weeks of the spring semester for 2 consecutive years. The control group participants ($n = 22$) were recruited from the same pool of traditional undergraduate students. These students were not enrolled in any specific course or engaged in any directed meditation or spiritual disciplines. Pre- and postmeasurements were gathered in the first 7 weeks of the fall semester with 4 weeks in between the measurements. Students from both groups were from a variety of majors, primarily social sciences, humanities, and religion. The age range was 17–42 with a mean age of 23 years. The median age was 21. Of the original 62 students completing the pretesting, two were omitted from analysis due to not returning for the second testing session. This research was approved by the Northwest Nazarene University institutional review board.

Materials

The *Beck Depression Inventory* II (BDI; Beck, Steer, & Brown, 1996) and the *State-Trait Anxiety Scale* (STAI; Spielberger, 1983) are self-report behavioral and mood measures used to establish similarities in mood states between the control and experimental groups. They were also used to evaluate changes across the first 5 weeks of the semester.

Students used a daily record sheet to record their sleep, quality of sleep, type of spiritual discipline, beginning and ending time of the discipline, any naps taken during the day, amount of caffeine consumed, and self-rating of mood. Students kept the records daily and turned them in each week.

All physiological measurements were gathered using the Biopac data acquisition system, M150. A two channel galvanic skin response (GSR) was used to measure microseman (sweat) changes during the rest and cognitive task pre- and postmeditation period. A two channel electrocardiogram (ECG) was used to record heart rate (BPM) during the rest and cognitive tasks for both pre- and posttesting. Millivolts were recorded from a 12 channel electroencephalogram (EEG) during the task for pre and post. Four frontal lobe (F3, F4, F7, and F8), two temporal lobe (T3, T4) and two parietal lobe (P3, P4) were recorded and analyzed. Two occipital lobe (O1, O2) were also recorded because we were using a visual task for stimulation. FP1 and FP2 were used to monitor possible interference from eyeblinks. The reference points were the participant's earlobes on which single gold electrodes were placed.

SuperLab 4.5 software (Biopac Systems Inc., 2010) was used for presentation of the rest stimulus, instructions, and visual cognitive task. A static picture of tulips was projected on the computer monitor for the rest period. The participant sat approximately 53 cm from the monitor. The keyboard was on the table between the participant and the monitor stand. For the rest phase, the picture was shown for 5 minutes. In Phase 1, color (e.g., blue, green, maroon) or noncolor words (e.g., truck, tree, pencil) were shown on a computer monitor at eye level of the participant. Emotion words (e.g., sad, disgusted, happy) were introduced during Trial 2 of the Color Word task but no specific instructions were given. A legend for key strokes was taped to the bottom of the monitor for reference. The words were randomized for each trial of three trials. The words were all in black ink, lower case letters and in 48-point font and centered on the monitor. For Phase 3, an adaptation of the Stroop task (obtained from Biopac) was used. There were five trials within the Stroop task: black color words (e.g., black, red, green), color xxx, colored words (read the word), colored words (say the ink color), and black color words. The words were shown on the screen until the participant pressed a key. Correct responses and response time were recorded for the Phase 2 task. Response time only was recorded for the Stroop task.

Acqknowledge EEG analysis software (Biopac Systems Inc., 2011) was used to analyze all the physiological recordings. SPSS was used to analyze the data.

Procedure

On the initial day of class, a research assistant invited students enrolled in the spiritual disciplines and neuroscience course to allow their data to be used in the study. Since students were using their pre- and postrecordings to write their paper for the course, it was considered important that the research assistants explain the procedure to students and gain consent while the professors were out of the classroom. The research assistants scheduled appointments for prerecordings with each student. All students in the course completed the BDI and STAI in class and turned them into the research assistants for scoring. Students arrived individually at their appointed time to the psychology lab for the physiological recordings. They were give a code and all data were recorded using the code. The master list was kept by the research assistants in a locked location.

During the rest phase, students were instructed to look at the picture on the monitor and relax. No further instructions on how to relax were provided. During the Color Word task, students were instructed to focus on the "+" in the middle of the screen and read the words that appeared. If the word was a color word the student was instructed to press the "c" key on the keyboard. If the word was a noncolor word, students were instructed to press the "n" key as quickly as possible. During Trial 2 of the Color Word tasks, emotion words appeared but students were not instructed to act on these words. The color words included more challenging words (e.g. russet, chartreuse). During Trial 3, students were instructed to continue pressing "c" for color words, "n" for noncolor words, and also "b" for emotion words. During the Stroop phase, students were instructed to press keys (z, x, c, v, b) that were representing each color on the legend taped to the monitor.

Students in the experimental group continued with the course. During each weekly class session, the spiritual disciplines were taught and practiced. These disciplines included meditation, silence and solitude, silent prayers, silent reading, and journaling. Students were encouraged to use one discipline each week rather than changing during the week. During the fourth week of class, the research assistants again set appointments for testing. The BDI and STAI were again completed in class. Students scored their own BDI and STAI during class guided by the professor. They turned in their scores to the research assistant. Research assistants analyzed the EEG, ECG, and GSR data. Individual scores were provided to each student for writing her or his reflection paper for the course. The preliminary findings were discussed during the final class session.

Students participating in the control group were recruited through announcements in psychology and religion courses. The research assistants set appointments for those who responded. These students signed consent forms and completed the BDI and STAI in the lab since they were not in a single course as a group. The physiological measures were taken using the same programs and recordings as the experimental group. At the end of the session, the next session was scheduled for 4 weeks out. Students were sent email reminders for the postsession. These students were invited to attend a presentation of the data that occurred in the spring semester.

Results

Psychological Measures

The data were analyzed using 2(group)X2(pre-post) repeated measures ANOVAs for each dependent variable (BDI score, State score, Trait score).

Pretesting: BDI and STAI. There is a significant difference in the BDI scores between groups before the intervention, $t(58) = 2.497$, $p = .015$, $Me = 8.99$, $Mc = 4.59$). Although statistically different, both of these values are well below the clinical cutoff score for the BDI, thus both groups are within the normal range for mood. There is no difference in the state anxiety levels between the experimental and control group prior to the intervention, $t(53,743) = 1.717$, $p > .05$, ($Me = 36.03$, $Mc = 31.41$). Both means were within the average range for college students. There is a significant difference in trait anxiety between the experimental and control groups prior to the intervention, $t(55.99) = 3.856$, $p = .0001$ ($Me = 41.49$, $Mc = 32.67$). The experimental group mean falls above one of the suggested cutoffs for anxiety.

BDI and STAI pre-post. There is a main effect for "time" for the BDI, $F(1,53) = 4.617$, $p = .036$. There was a significant decrease in BDI scores (the full group) at the posttesting, $Mpre = 7.02$, $Mpost = 5.6$. There is no significant interaction between the groups and the times for the BDI, $F(1, 53) = .042$, $p < .05$. There is no main effect for time with state or trait anxiety scores. There is no significant interaction between time and group for state or trait anxiety scores.

Physiological Measures

The data were analyzed using 2(group)X2(pre-post)X3(phase) ANOVAs for GSR values and BPM values.

GSR-Rest. There was a trend for time on GSR measurements, $F(1,55) = 3.343$, $p = .073$. The GSR ulms were significantly higher during the post-reading ($Mpost = 5.27$) than the preintervention reading ($Mpre = 4.12$). There was a significant interaction between time and groups for GSR, $F(1,55) = 11.071$, $p = .002$. The experimental group GSR decreased after intervention whereas the control group increased (See Figure 11.1).

GSR – Color. There is a main effect for time for GSR ulms, $F(1,55) = 5.133$, $p = .027$, ($Mpre = 5.41$, $Mpost = 7.18$). The ulms increased after the 4 weeks. There is a significant interaction for GSR between time and group, $F(1,55) = 10.710$, $p = .002$ (See Figure 11.2).

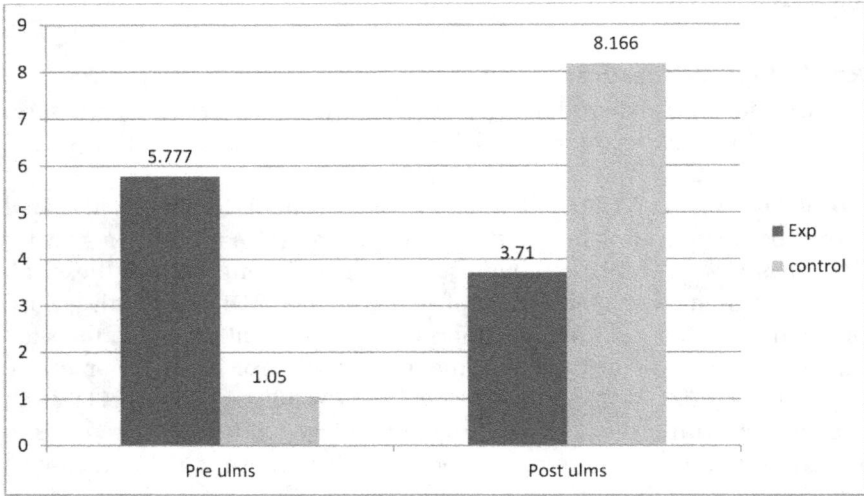

Note: The GSR measurements (uhms) for the experimental group started out higher than for the control group and decreased after the 4-week intervention of using spiritual disciplines daily. The GSR for the control group started out lower but increased significantly following the 4 weeks between pre- and postmeasurements.

Figure 11.1.

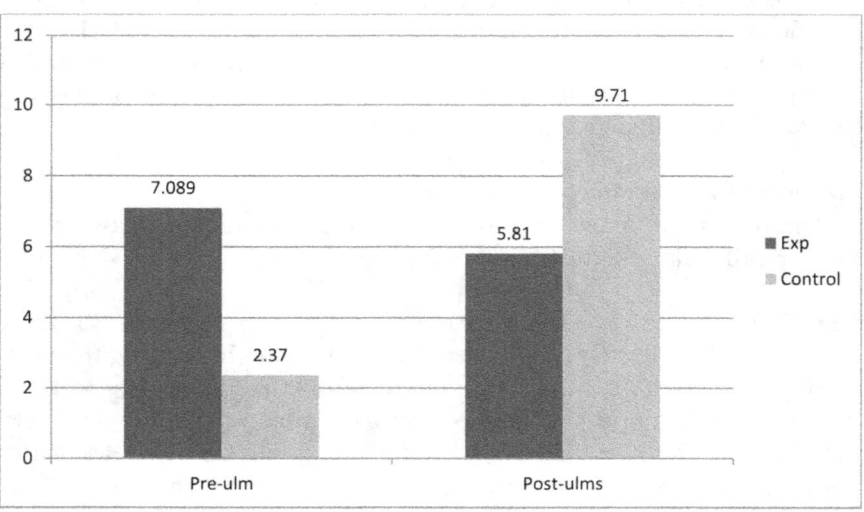

Note: The GSR ulms decreased significantly during the color word task for the experimental group after daily spiritual disciplines and increased significantly for the control group after the 4 weeks. This indicates that the intervention of using spiritual disciplines daily decreased anxiety as the student progressed through the semester.

Figure 11.2.

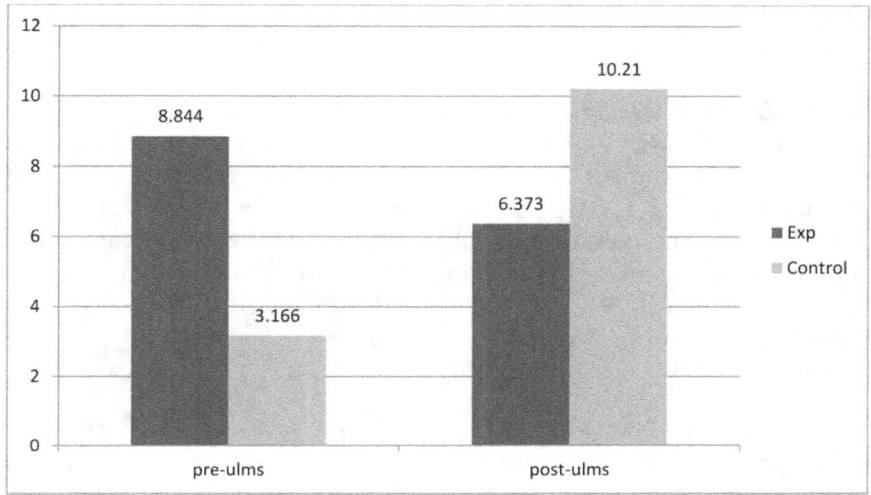

Notes: Those in the experimental group significantly decreased their GSR measures after 4 weeks of spiritual disciplines. Those in the control group had a significant increase in GSR ulms over the 4 weeks of academics only.

Figure 11.3. GSR-Stroop.

GSR—Stroop. There is no main effect for time. There is a significant interaction for the Stroop GSR measures, $F(1,55) = 7.507$, $p = .008$ (See Figure 11.3).

Heart Rate (BPM)-Rest. There is a main effect for time, $F(1, 53) = 8.487$, $p = .005$. There was a significant decrease in BPM during the second testing ($M_{pre} = 79.24$, $M_{post} = 73.24$). There was no significant interaction between time and group and no group difference in heart rate.

BPM-Color. There is a main effect for time, $F(1, 53) = 7.100$, $p = .01$. There was a significant decrease in BPM during the second testing ($M_{pre} = 81.998$, $M_{post} = 76.092$). There was no significant interaction between time and group and no group difference in heart rate.

BPM-Stroop. There is a main effect for time, $F(1, 53) = 7.793$, $p = .007$. There was a significant decrease in BPM during the second testing ($M_{pre} = 83.297$, $M_{post} = 77.705$). There was no significant interaction between time and group and no group difference in heart rate.

Electroencephalogram (EEG)

The mean power data were adjusted using syntax in SPSS in order to be able to conduct the analysis. A 2(group)X2(pre-post)X3(phases) MANOVA was completed using each electrode area measured.

Temporal Lobe. No main effects or interactions occurred in the left or right temporal lobe.

Frontal Lobe Left (Medial). There is a main effect for trial, $F(1, 54) = 6.055$, $p = .017$ ($Mpre = 9.09883\text{E-7mv}$ $Mpost = 6.3659\text{E-7mv}$). There was more activity in the left frontal during the posttesting. There is a main effect for group, $F(1, 54) = 11.609$, $p = .001$ ($Me = 4.368\text{E-7mv}$, $Mc = 1.425\text{E-6mv}$). The control group had higher levels of electrical activity than the experimental group. There is a significant interaction between group and trial, $F(1, 54) = 12.374$, $p = .001$ (See Figure 11.4). There is a significant three-way interaction between group, trial and phase, $F(2, 108) = 3.472$, $p = .037$ (See Figure 11.5).

Frontal Lobe Right (Medial). There is a main effect for trial $F(1,53) = 4.686$, $p = .035$ ($Mpre = 2.97327\text{E-6}$, $Mpost = 2.107\text{E-7}$). There was less electrical activity at the posttest for the full group.

There is a main effect for condition, $F(2, 106) = 7.427$, $p = .001$ (Rest = .000002163, Color = .0000023747, Stroop = .0000030279). There was

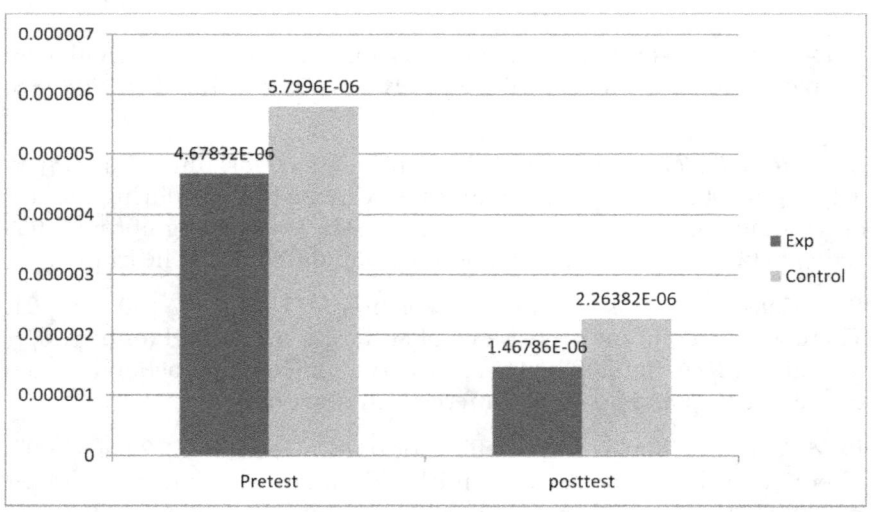

Note: Both groups show a decrease in Mv mean power in left frontal activity with the post-testing. The experimental group change is significant. Lower mean power while able to continue accurate cognitive activity suggests less energy is necessary to complete the activity. Both groups show less energy probably due to familiarity with the task. The experimental group change is a significant difference suggesting not only familiarity with the task but possibly the effects of using spiritual disciplines to enhance overall cognitive functioning.

Figure 11.4.

Changing Behavior and Renewing the Brain 147

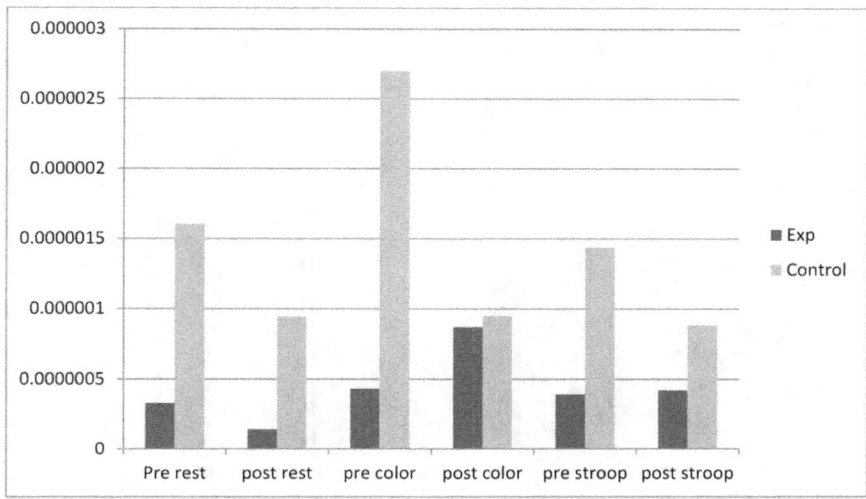

Note: The experimental group exhibited less electrical activity than the control group across all conditions except postcolor. Looking at patterns within the groups, there is a decrease for the control group for all conditions. This may be a learning effect. The exception is with the color (light cognitive load) for the experimental group. This group increases the electrical activity following the 4 weeks of spiritual discipline. There is no change in activity for the Stroop for the experimental group.

Figure 11.5.   Left frontal trial x group x phase.

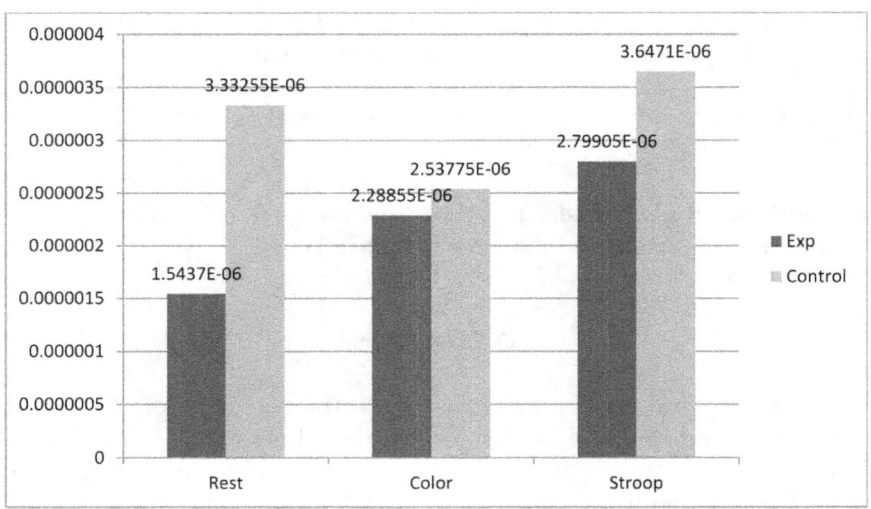

Note: The experimental group increases electrical activity for each additional cognitive load (condition). The control group shows a decrease for the light cognitive load.

Figure 11.6.   Right frontal (F4) group X condition.

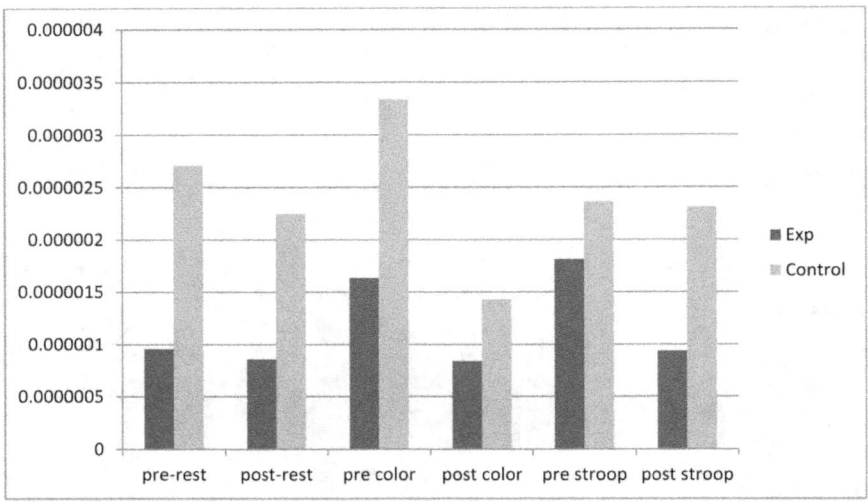

Note: More electrical energy is used by the control group for both pre and post. Looking at the patterns of changes between pre and post, the control decreased electrical activity for the rest and color conditions but not the stroop. The experimental group decreased electrical activity for the higher cognitive functions of color and Stroop conditions.

Figure 11.7. Right parietal trial X condition X group.

increased activity with each condition for the full group at both pre- and posttesting. There is a significant interaction for condition X group, $F(2, 106) = 2.993$, $p = .054$. There were no main effects or significant interactions for the right or left lateral frontal measurements (F7, F8).

Parietal Lobe Left. There is a main effect for trial $F(1, 53) = 4.516$, $p = .038$.

Parietal Lobe Right. There is a three-way interaction for trial X condition X group $F(2,108) = 3.296$, $p = .041$ (See Figure 11.7).

DISCUSSION

Can psychological and physiological measurements be altered within 4 weeks by introducing the use of spiritual disciplines for 20 minutes each day for just a month? That was the question asked at the beginning of the course. Neuroscientists have demonstrated that brainwaves and heart rate can be changed when a person is in a state of meditation (Newberg & Waldman, 2009). Also, studies show that a person trained by a meditation expert or guru can see physiological changes (Newberg & Waldman,

2009). Can these disciplines affect the psychological well-being and physiological status, including brain energy, when done by a person individually for merely 20 minutes per day? The study shows a resounding *yes*.

After presenting the preliminary results from the first experimental group, it was suggested that maybe just being in college and taking courses through the first month of the semester was the reason for the changes we found. The research shows that this is not accurate. None of the psychological or physiological measures improved for the control group alone. The BPM decreased between pretesting and posttesting for both groups. It is possible that this change was due to the initial novelty of having the electrodes placed and proceeding through the first testing. Novelty and familiarity does not explain the improved mood of the students in the experimental group as measured by the BDI. It is possible that mood symptoms were improved due to the experience of the course rather than the spiritual disciplines or a combination of the course, and spiritual disciplines may have resulted in the significant increase in mood.

The results of the GSR provide some additional support for the positive effects of daily spiritual disciplines. Even with the increase in difficulty of the stimulus task, the GSR of the experimental group decreased and the control group increased in the posttest measurement. If general anxiety has decreased and well-being has increased due to the use of daily centering and focus on spiritual matters, it makes sense that the GSR would reflect this positive effect.

The increased mean power demonstrated especially in the right frontal lobe is consistent with the increasing cognitive load from rest to simple response to color words to the complex inhibition task of the Stroop. This increase in effort was evident for the full group both at the preintervention and postintervention testing.

Although brain functioning is interconnected, there are some tasks that seem to rely on one area of the brain more than others. For example, the language center is basically a left hemisphere functioning for the majority of people (Gazzaniga, Ivry, & Mangun, 2002). A person's memory involves areas of both hemispheres beginning with working memory in the frontal lobes (Baddeley, 1986, 2010), yet memory for verbal information tends to be performed in the left hemisphere and memory for visual information tends to rely on the right hemisphere. With tasks that can become automated (e.g., performances), humans begin learning using the more sequential left hemisphere and then need less energy after the last is learned. When a person is feeling anxious or struggling with high levels of stress, cognitive tasks become more challenging because the response to anxiety and stress appear to cause us to think much more attuned to a "beginner" using more of the sequential left hemisphere and the frontal lobe. This would lead to higher levels of energy from the left hemisphere. As anxiety or stress response decreases, the same cognitive task should require less energy and thus the Mv from the EEG would decrease.

The parietal lobe is partially responsible for the ability to attend (Gazzaniga, 2004; Yin et al., 2012). The experimental group had decreased energy in the right parietal for the two cognitive tasks (color word and Stroop), whereas the control group decreased only for color word. Since the accuracy of the responses did not decrease on the posttest trial, it might suggest that the influence of using spiritual disciplines lead to more efficient use of energy for focusing. Thus the mind was less distracted due to the use of spiritual disciplines and less cell energy was needed in order to attend to the cognitive tasks including the heavy and complex cognitive load of the Stroop task.

There are several questions about meditation and brain waves that need to be answered by further research, but what is amazing from this study is that brain waves changes occurred, depressive symptoms decreased, "anxiety" as measured by galvanic skin response decreased, and students reported positive changes in sleep, attention, concentration, and self-efficacy. These outcomes occurred with a mere 20 minutes of meditation each day within 4 weeks. The take away message is that we can positively change our brain activity with a minor change in behavior. Spiritual disciplines not only focus us toward our relationship with God, but in doing so also positively affect our mood and improve our cognitive abilities. It would be interesting to see what changes occur if these practices were measured after 6 months, 1 year, 5 years, or 25 years of practice.

DISCUSSION QUESTIONS

1. What is neurotheology? What benefits and challenges of this new area of study bring to Christian religious experiences?
2. In what ways does neuroscience affirm aspects of Christian spirituality, particularly the role of prayer and meditation on overall well-being?
3. What are examples of inward and contemplative practices?
4. What are some spiritual practices that I can include in my life to improve my overall well-being?

REFERENCES

Acqknowledge Software. (2011). Camino Goleta, CA: Biopac Systems, Inc.
Baddeley, A. (1986). *Working memory*. New York, NY: Oxford University Press.

Baddeley, A. (2010). Working memory. *Current Biology, 20*(4), R136–R140.
Beck, A., Steer, R., & Brown, G. (1996). *Beck depression inventory–II*. San Antonio, TX: Pearson Education.
Begley, S. (2007). *Train your mind, change your brain: How a new science reveals our extraordinary potential to transform ourselves*. New York, NY: Ballantine Books.
Calhoun, A. C. (2005). *Spiritual disciplines handbook: Practices that transform us*. Downers Grove, IL: InterVarsity Press.
Gazzaniga, M. (2004). *The cognitive neurosciences*. Cambridge, MA, MIT Press.
Gazzaniga, M., Ivry, R., & Mangun, G. (2002) *Cognitive neuroscience: The biology of the mind*. New York, NY: Norton.
Jeeves, M., & Brown, W. S. (2009). *Neuroscience, psychology and religion: Illusions, delusions, and realities about human nature*. West Conshohocken, PA: Templeton Foundation Press.
Jones, T. (2005). *The sacred way: Spiritual practices for everyday life*. Grand Rapids, MI: Zondervan/Youth Specialties.
Leclerc, D., & Maddix, M. A. (2011). *Spiritual formation: A Wesleyan paradigm*. Kansas City, MO: Beacon Hill Press.
Newberg, A., & Waldman, M. R. (2010). *How God changes your mind: Breakthrough findings from a leading neuroscientist*. New York, NY: Ballantine Books.
Spielberger, C. (1983). *State-trait anxiety inventory*. Menlo Park, CA: Mind Garden.
SuperLab System. (2010). Camino Goleta, CA: Biopac Systems.
Yin, X., Zhao, L., Xu, J., Evans, A. C., Fan, L., Ge, H., ... De Lange, F. (2012). Anatomical substrates of the alerting, orienting and executive control components of attention: Focus on the posterior parietal lobe. *PLoS ONE, 7*(11), e50590. doi:10.1371/journal.pone.0050590

SUGGESTED FOR FURTHER READING

Barrett, J. L. (2011). *Cognitive science, religion, and theology: From human minds to divine minds*. West Conshohocken, PA: Templeton Foundation Press.
Jeeves, M., & Brown, W. S. (2009). *Neuroscience, psychology and religion: Illusions, delusions, and realities about human nature*. West Conshohocken, PA: Templeton Foundation Press.
Jones, T. (2005). *The sacred way: Spiritual practices for everyday life*. Grand Rapids, MI: Zondervan/Youth Specialties.
Leclerc, D., & Maddix, M. A. (2011). *Spiritual formation: A Wesleyan paradigm*. Kansas City, MO: Beacon Hill Press.
McNamara, P. (2009). *The neuroscience of religious experience*. New York, NY: Cambridge University Press.
Newberg, A., & Waldman, M. R. (2009). *How God changes your mind: Breakthrough findings from a leading neuroscientist*. New York, NY: Ballantine Books.

CHAPTER 12

EQUIPPING MINDS FOR CHRISTIAN EDUCATION OR LEARNING FROM NEUROSCIENCE FOR CHRISTIAN EDUCATORS

Carol T. Brown
Executive Director, Equipping Minds

INTRODUCTION

Over the last 20 years, research on working memory found reliable correlations between working memory span and several other measures of cognitive function, intelligence, and performance in school (Alloway, 2011). The key to intelligence is being able to put those facts together, prioritize the information, and do something constructive with it. The "working memory capacity refers to the ability to hold information in mind while maintaining other information to achieve a cognitive task" (Camos, 2008, p. 38). Working memory is the skill that gives a person the advantage of managing all this information and is a stronger indicator of a learner's academic and personal potential than an IQ test (Alloway & Alloway, 2014).

The discoveries in neuroscience, in conversation with the theories, programs, and research of Dr. Reuven Feuerstein bring hope to people working with learners possessing neurodevelopmental learning disorders (NLD): autism spectrum disorders, attention deficit hyperactivity disorder (ADHD), specific learning disorder, intellectual disability (Intellectual Developmental Disorder), communication disorders, and motor disorders. Dr. Feuerstein is a clinical and cognitive psychologist who has shown that cognitive functioning is modifiable through mediated learning interventions (Feuerstein, Falik, & Feuerstein, 2015). Parents, teachers, and interventionists need to be informed and equipped with the methods and tools to improve a learner's cognitive abilities rather than focusing on remediation of subject content alone. New concepts of a learner's ability and development are needed.

EDUCATIONAL APPROACHES

Christian educators have come to accept the theories of human development embraced by the American educational system that discount spirituality and have a naturalist worldview. These developmental theories inform our curricula, determine who may or may not attend Christian schools, define what is normal, and identify one's cognitive potential based on an intelligence quotient (IQ), a static assessment (Feuerstein, Feuerstein, & Falik, 2010). Educational psychology and human and child development textbooks have been the primary guide for understanding learners and have historically begun with Piaget's theory on cognitive development (Ormrod, 2010). Piaget believed every learner was responsible for generating his or her own "logical structures." The progression and acquisition of these abilities resulted from a learner's successful interactions with the environment (Piaget, 1952). This belief system led many to view learners with neurodevelopmental learning disorders as having a fixed limit to their cognitive abilities since they were not able to acquire these abilities on their own. This belief led to the different approaches for learners with developmental disorders.

The first group of educators follows a traditional approach where one finds those who believe in full integration of all children in typical schools and classrooms, usually with an individual aide. The curriculum is adapted to suit the individual, as academics are a secondary focus, which excludes reading instruction beyond a basic functional level. The primary focus is on social adaptation and vocational and daily living skills (Nevin, 2000).

The second group of educators follows a progressive approach and believes that academic skills such as arithmetic and reading remain crucial to survive and sees social skills as a secondary emphasis. This group tends to prefer specialized schools or separate classes within typical schools. Many researchers and educators combine the two approaches in different ways (Katims, 2000). Julie Lane and Quentin Kinnison (2014) follow a combined approach by informing Christian schools on the policies and procedures for developing a special needs program. The schools are encouraged to follow the public school model that focuses on remediation, accommodation, modification, and intervention.

There have been advances in effectively including learners with NLD in terms of educational policy, philosophy, and curriculum. Numerous researchers have studied the development of reading and mathematical skills in learners with learning disabilities. However, the cognitive enhancement of learners with severe NLD receives inadequate attention (Kozulin et al., 2010). Yet the research done substantiates that learners with intellectual disorders can participate and benefit from cognitive development and enrichment programs. The "Bright Start" program of Brooks and Haywood, which is based on Feuerstein's theories, increases intelligence quotient (IQ), enhances logical reasoning and problem-solving skills, allows children to be included in the regular classroom, and increases academic performance and intrinsic motivation (Haywood, 2004). Paour's (1993) "transformation box" program and Klauer's (2002) inductive reasoning program have demonstrated the ability of learners with intellectual disorders to move beyond the preoperational level of thinking.

In response to these developments, a third group merges the goals of both of the above groups by emphasizing neither of those basic approaches but rather the idea of cognitive development or cognitive education as a goal in itself. A learner's social and intellectual development is interrelated (Feuerstein & Rand, 1997). The teacher is a mediator who invites the learner to identify a problem, to analyze it, to use inductive thinking processes to develop a strategy for its solution, and to connect it to other knowledge networks. Teachers who apply these principles of mediation enable learners to find a greater level of success as independent and active students (Feuerstein et al., 2015).

NEUROSCIENCE CONFIRMS THE BRAIN CAN CHANGE

The belief that cognitive abilities are fixed and nonmodifiable has been prevalent in the United States for many years (Tan & Seng, 2008). An individual's intellectual ability has been measured by their "intelligence quotient" (IQ) (Patel, Aronson, & Divan, 2013). Proponents of this *fixist* point of view believe that change in functioning and behavior cannot be made beyond a certain level (Sternberg, 1984). Over the last two decades the field of neuroscience has used noninvasive technologies, such as the fMRI and PET, to show the plasticity of the brain, or *neuroplasticity*, which is the brain's ability to heal, grow, and change (Boleyn-Fitzgerald, 2010). These imaging techniques show brain activity during development and learning. "It is now increasingly recognized that the brain is not a static structure and is in fact a modifiable system that changes its physical and functional architecture in response to its complex interaction with its internal processes and the environment" (Tan & Seng, 2008, p. ix).

According to Boleyn-Fitzgerald (2010), research confirms the modifiability of the brain through experience and training (Boleyn-Fitzgerald, 2010). Nobel Prize winner Eric Kandel demonstrated that learning can ignite genes that change neural structure. Norman Doidge (2015) recounts the current advances of international scientists who demonstrate that the brain is plastic. Recognizing that intelligence can be improved, educators can adapt their teaching to include students traditionally seen as difficult to teach such as learners with (NLD): autism spectrum disorders, attention deficit hyperactivity disorder (ADHD), specific learning disorder, intellectual disability (Intellectual Developmental Disorder), communication disorders, and motor disorders. If intelligence is constantly changing, or even potentially changing, the Christian educator must embrace these families and students who are created in the image of God for His purposes and His glory.

REUVEN FEUERSTEIN: PIONEER OF NEUROPLASTICITY

The first program to increase intellectual performance with learners with neurodevelopmental learning disorders was developed more than 50 years ago by Reuven Feuerstein. As a clinical and cognitive psychologist he believed that intelligence was changeable and modifiable regardless of age, genetics, neurodevelopmental conditions, and developmental disabilities (Feuerstein et al., 2010). Feuerstein worked with a wide range of different groups of people—from Holocaust survivors, to people who had suffered from brain damage, down syndrome, and autism, to those who are intellectually gifted. When he began working with the children who had survived the Holocaust, the goal was to rehabilitate them from their traumatic experiences (Feuerstein et al., 2010).

Feuerstein also disagreed with the accepted concepts of the *critical period* or *critical age*, which states that if a person has not reached a particular function by a certain age, he or she no longer has the ability to learn that skill. According to Brian Boyd (2007), in speaking of the two components of modifiable intelligence, the intellect and the emotion, Feuerstein would begin with an unusual perspective, an expression of faith:

> But the point we wish to emphasize is that in the beginning there must be a need—a need that will generate the belief in human modifiability. I must have the need to have my students and those with who I am engaged reach higher potentials of functioning. This need energizes me to act and motivates my faith (belief) that there are positive, effective, and meaningful alternatives to be found, to fight for, and to bring this faith into being. I believe that the student is a modifiable being who is capable of change and capable of changing according to his or her will and decisions. Human beings' modifiability differentiates them from other creatures and, according to the Rabbinic Midrash, "even from the angels." Herein lies the main uniqueness of human beings. (Feuerstein et al., 2010, p. 6)

"Belief in modifiability" provides an essential element of Feuerstein's Theory of Structural Cognitive Modifiability (Feuerstein et al., 2010).

INTELLIGENCE IS MODIFIABLE

Since the 1950s, Feuerstein observed the modifiability of the brain through the application of MLE. Today the discoveries in neuroscience confirm and support Feuerstein's theory known as Structural Cognitive Modifiability (SCM) that presents an optimistic view of the learner and one's propensity to be modified. Feuerstein's theory of human development includes three basic ideas:

1. Three forces shape human beings: environment, human biology, and mediation.
2. Temporary states determine behavior: How someone behaves—namely emotional, intellectual, and even habitually learned activities—represents a temporary state, not a permanent trait. This means that intelligence is adaptive. In other words, intelligence can change; it is not fixed once and for all.
3. The brain is plastic: Because all behaviors are open and developing, the brain can generate new structures through a combination of external and internal factors (Feuerstein, Feuerstein, Falik, & Rand, 2006).

Feuerstein insisted that human cognitive abilities can be changed regardless of etiology, severity, or a person's age, even if the condition is generally considered irrevocable and irreparable. "Don't tell me what a person is," said Feuerstein. "Tell me how he is changeable" (Feuerstein & Lewin-Benham, 2012, p. 30).

LEARNING THROUGH MEDIATION

The theory of mediated learning experience (MLE) initially grew as part of Feuerstein's theory of structural cognitive modifiability (SCM). Mediation describes an interaction in which a mediator who possesses knowledge conveys a particular meaning or skill to a child and encourages him or her to transcend, that is, to relate the meaning to some other thought or experience. Mediation is intended to help children expand their cognitive capacity, especially when ideas are new or challenging. Piaget advocated for a natural progression of learning through direct exposure to stimuli, or the "stimulus-organism-response (S-O-R)" model, which holds that it is enough for a person to simply dialogue with nature and the environment for cognitive development to occur (Feuerstein et al., 2015). Feuerstein believes a human mediator is needed, or "stimulus-human-organism-human-response (S-H-O-H-R)," allowing the mediator to take the learner beyond the natural limitations to reaching his or her full cognitive potential and generate new cognitive structures (Feuerstein et al., 2015).

While Piaget and Feuerstein are both giants in the field of human development, the greatest differences are their beliefs in fixed versus changeable intelligence and the role of a human mediator in developing a child's intelligence (Feuerstein et al., 2010). Piaget did not believe that adults are any different from other objects that provide information, and thus they should not intervene in a child's activity. He believed in spontaneous development: "I will call it psychological—the development of the intelligence itself, what the child learns by himself, what none can teach him, and what he must discover alone" (Piaget, 1973, p. 2). Feuerstein, however, sees the human mediator as essential for an individual to grow (Feuerstein et al., 2010). Feuerstein has sought to identify and correct these deficits to enable students to reach their full cognitive potential, as well as to increase their internal motivation and personal confidence. By using mediation, these deficient functions can be formed and modified in significant ways (Feuerstein et al., 2010).

Instrumental Enrichment

The theory of structural cognitive modifiability (SCM) and the applications of the mediated learning experience (MLE) are the foundation of

Feuerstein's Instrumental Enrichment (FIE) standard and basic programs that were developed over 40 years ago. FIE is a cognitive development program emphasizing critical thinking strategies. Fourteen instruments are designed to build the perquisites and processes of learning rather than academic content or skills. They can be implemented in a classroom or as a therapeutic intervention in a small group or an individualized basis. FIE initially focused on culturally deprived and low-functioning children and adolescents with chromosomally determined conditions to build their cognitive functions and structures. The program has expanded to include learners of all ages and abilities to strengthen their learning capacity (Feuerstein et al., 2006).

Cognitive Functions

Reuven Feuerstein defines cognitive functions as "thinking abilities" that can be taught, learned, and developed. Feuerstein has categorized the cognitive functions according to the three major phases of the mental act: input, elaboration, and output. Although artificially separated into three phases, they don't necessarily occur separately in life. However, the subdivision is useful to analyze and describe thinking as well as to determine what factors might negatively affect thinking. Teachers and parents can use this model to better understand and help the child who is experiencing difficulties with a particular task. By having a working knowledge of the cognitive functions, teachers (mediators) can differentiate errors due to a lack of knowledge or from a deficient cognitive function (Feuerstein et al., 2006). For example, if a child fails in the task of classification, it is not enough to comment on the child's poor intelligence or inability to classify, but rather the underlying causes of the difficulty (which can be found in one of the three phases of thinking) should be sought. The inability to classify, for instance, may be due to underlying underdeveloped functions, such as imprecise data gathering at the input phase or poor communication skills at the output phase. A detailed analysis of a student's cognitive functions requires an in-depth understanding of the three phases of the mental act (Mentis, Dunn-Bernstein, Mentis, & Skuy, 2009).

Research Studies on Cognitive Enhancement

The Feuerstein Institute has conducted research for the last five decades that confirms that cognitive abilities can be modified (Tan & Seng, 2008). Instrumental Enrichment (FIE) and MLE have been found to have positive effects on many types of learners, including neurodevelopmental learning disorders (Kozulin et al., 2010). Many of these learners also have cultural deprivation and differences. These studies have encompassed many types of student populations using FIE (Feuerstein et al.,

2006) including attention deficit disorders, Autism, learning disabilities, and developmental disabilities.

In 2014, Krisztina Bohács's PhD thesis on clinical applications of Feurstein's mediated learning studied learners from 2 to 14 years of age with mild to moderate intellectual developmental disorders: genetic syndromes, cerebral paresis, ADHD, and Autism. The Raven Colored Matrices showed an increase in general intelligence, and there were significant changes in the cognitive development. There was also growth in domains necessary for school readiness. Bohács concludes, "If applied systematically with children with intellectual disabilities for a longer period of time (maybe even for 3–4 years) the applied systems are expected to lead to increased learning effectiveness, more effective basic cognitive processes and thinking skills, and to prepare children for school learning and a better adaptation to the challenges of everyday life" (2014, p.18)

Four-Year Case Study With *Equipping Minds Cognitive Development Curriculum*

Bohács recommended a 3- to 4-year study with learners with NLD. An individual case study was done with the *Equipping Minds Cognitive Development Curriculum* (EMCDC) from 2010–2015 with a learner with a NLD, Down syndrome. In September 2010, Marie's parents contacted the researcher to discuss using EMCDC to strengthen Marie's cognitive abilities; visual and auditory processing speed, comprehension, working memory, long term memory, and reasoning skills. According to Marie's parents, despite all the support from Marie's teachers, occupational therapist, speech therapist, special education teacher, and principal in third grade, her Measures of Academic Progress (MAP) scores—yearly academic tests that measure student growth from semester to semester—stayed stagnant for a full year. In the fall of fourth grade the first MAP scores again showed no growth. The researcher reviewed the academic and psychological testing showing an intellectual disability with deficits in processing, working memory, comprehension, and perceptual reasoning. The researcher agreed to begin working with Marie using EMCDC.

With the support of the school system, the researcher worked with Marie an hour of every school day for the next 12 weeks. At the end of 9 weeks, the principal met the researcher at the school door and reported that Marie had increased 20 points in reading, 11 points in math, 25 points in science, and 17 points in language arts. These gains were unprecedented as students typically increase 3–5 points.

Until this time, Marie had made minimal progress and her academic test scores had remained static from third to fourth grade. The change in these scores had been achieved over the last 9 weeks through one-on-one cognitive developmental exercises for enhancing processing, working memory, comprehension, and reasoning, which was divorced from academic content. Previously, she had received the standard interventions: remediation of content, learning strategies, and accommodations. These may have short-term benefits, but were not targeting the underlying cognitive deficits in processing and working memory, which would increase her cognitive abilities.

Marie's progress proves significant for those who still believe 85% of the measureable intelligence is due to nature or one's genetic factors and only 15% due to nurture or environmental factors which holds to a limited potential for change. Since Marie has an intellectual disability and Down syndrome, many educators believe these disorders limit her ability for significant academic gains. However, Marie's improvement implies that cognitive developmental exercises do have far transfer effects to academic achievement for learners who have an intellectual developmental disorder. Below are the results of the MAP tests after that first 9 weeks and over the next 4 years. Figures 12.3, 12.4, 12.5, and 12.6 illustrate the MAP test results which demonstrate significant gains in academic abilities.

Marie would continue the cognitive developmental exercises and continue to progress academically for the next 4 years. Her Kentucky Performance Rating for Educational (KPREP) scores are illustrated on Figures 12.5, 12.6, and 12.7 showing gains in math, reading, and writing on-demand. Marie's student growth percentile (SGP) in reading was 93% in sixth grade and 7% in seventh grade. Her SCP was 63% in math as a sixth

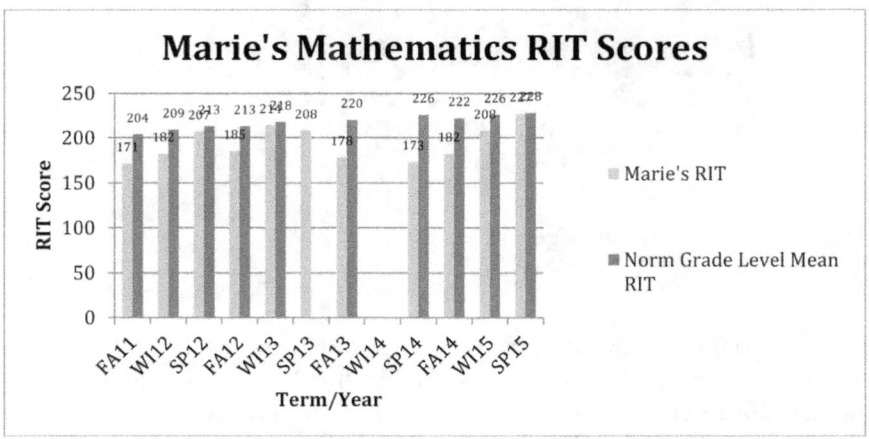

Figure 12.1.   Marie's mathematics RIT scores.

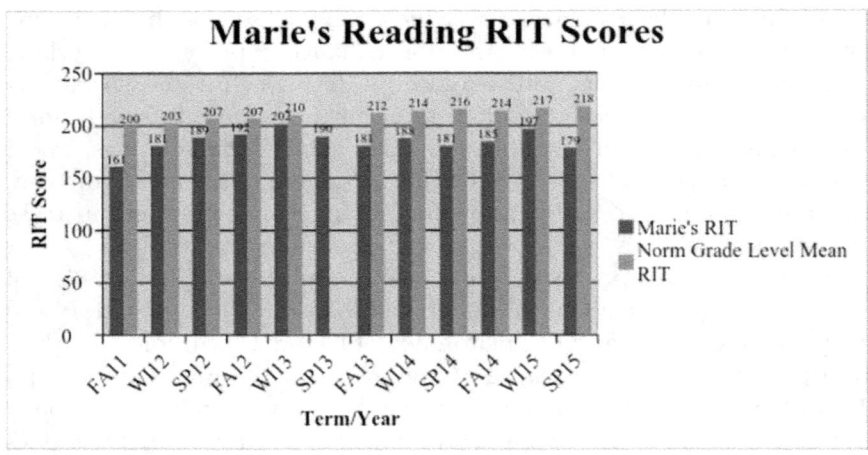

Figure 12.2. Marie's reading RIT scores.

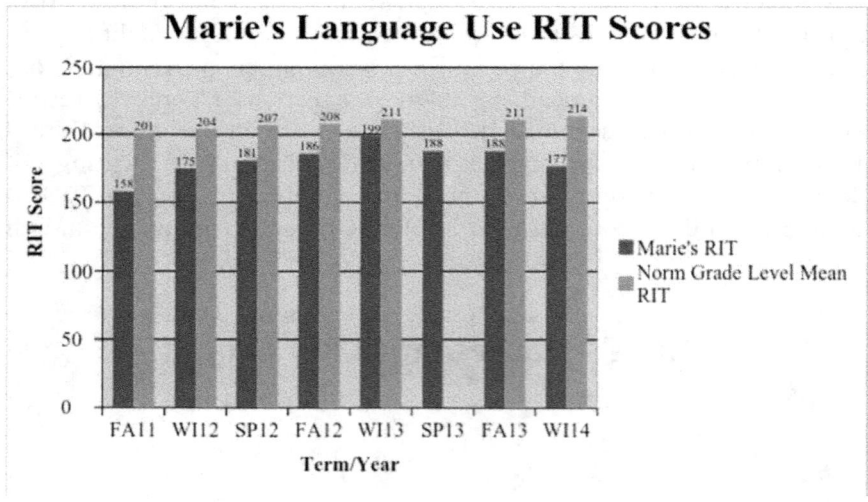

Figure 12.3. Marie's language RIT scores.

grader and 93% in seventh grade. Figure 12.8 illustrates the SCP for sixth and seventh grade. In 2015, she scored in the 39th percentile in mathematics, 36th percentile in science, and the 7th percentile in reading on the Stanford 10 National Assessment Ranking as a seventh grader. Figure 12.9 illustrates the Stanford National Ranking.

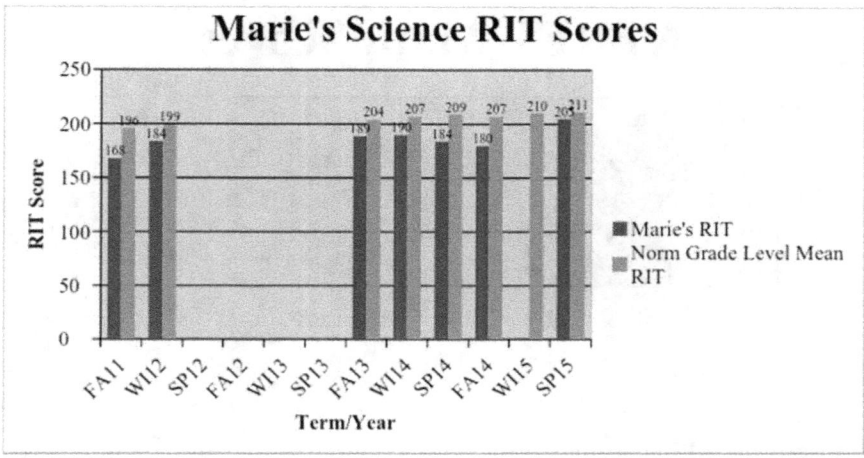

Figure 12.4. Marie's science RIT scores.

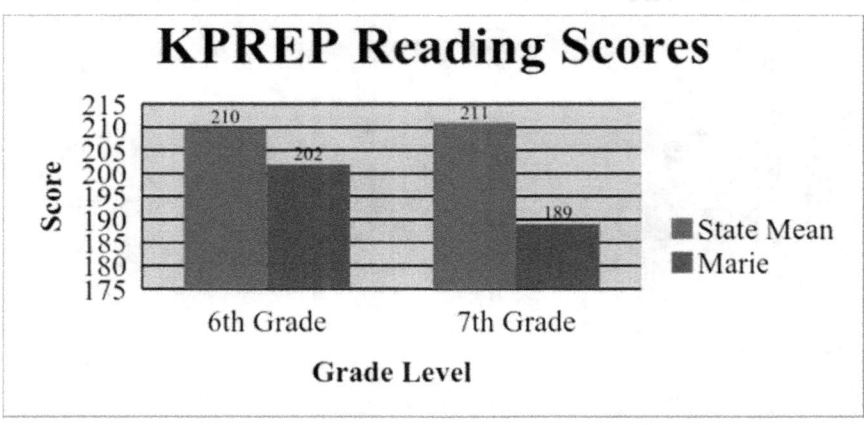

Figure 12.5. KPREP reading scores.

In conclusion, Marie's success may be attributed not only to supportive teachers, but undeniably to specific cognitive training exercises in EMCDC that are targeted to her areas of weakness. This past year Marie has been working more on logic, reasoning, and abstract thinking which is impacting her cognitive, social, and spiritual development. In the fall of 2015, Marie was baptized at her church after asking to meet with her pastor to discuss her relationship to Christ and desire to live for Him.

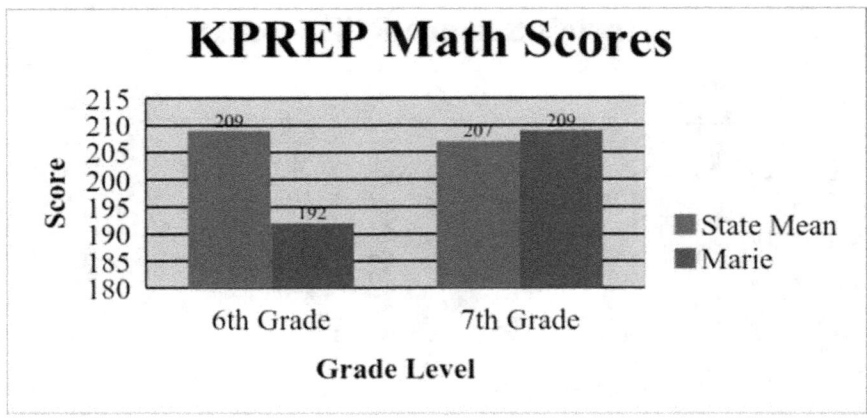

Figure 12.6. KPREP math scores.

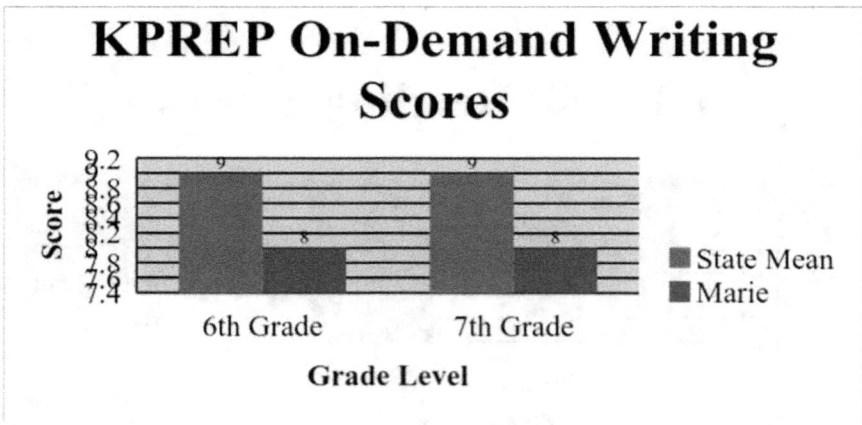

Figure 12.7. KPREP on-demand writing scores.

Some typical children and certainly those with special needs must have someone to "teach" their brains how to think, how to process information, and how to store information in the same way that children with special needs may need physical therapy to teach them how to roll over as infants or how to put one foot in front of the other to walk. The specific cognitive exercise Marie performs with Equipping Minds does just that for her brain.

The other important thing to note is that the educational director is continually changing the exercises as they are mastered and adapting the

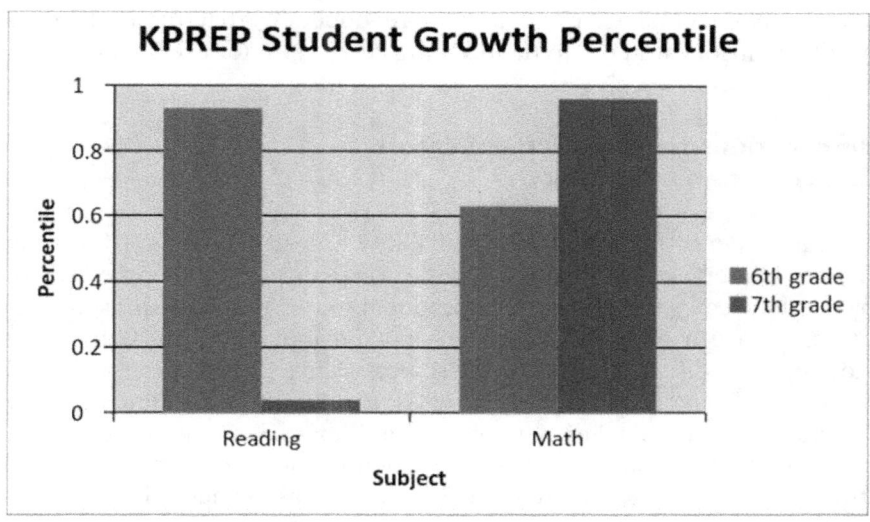

Figure 12.8. KPREP student growth percentile.

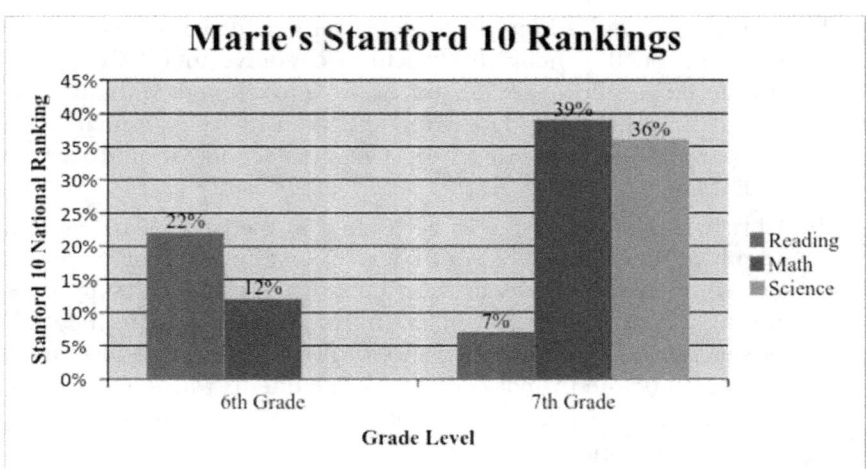

Figure 12.9. Marie's Stanford 10 rankings.

program for Marie. It's this individualized targeting of cognitive areas that sets this program apart from other programs. Just think, if this program can help a child with Down syndrome learn at this rate, imagine how it could help other children with neurodevelopmental learning disorders. Minimum remedial tutoring could be replaced with specific cognitive developmental exercises and more hopefully that these exercises would be incorporated into the teaching curriculum for every child.

Implications for Christian Educators in Schools and the Church

The evidence for cognitive modifiability in learners with NDL can no longer be denied. Christian educator's acceptance of development theories embraced by the American educational system that discount spirituality and have a naturalist worldview can be replaced with a theory of cognitive modifiability from a theist perspective. These developmental theories inform our academic and religious curricula, determine who may or may not attend Christian schools and participate in programs at church, define what is normal, and identify one's cognitive potential based on an intelligence quotient (IQ), a static assessment. The implications for the Christian school, the educator, and the church are substantial since intelligence can be developed when a mediator teaches and trains a student.

1. Christian school administrators, teachers, and parents should be educated on the theory of structural cognitive modifiability and how to be an effective mediator of the environment without overstimulating the child. The primary responsibility is ultimately on the parents and the church, and Christian schools should partner with them.
2. Christian educators need to be trained in mediated learning and cognitive developmental exercises. A combination of cognitive developmental exercises and curricular studies should result in significant advancement of both cognitive and domain-specific skills of special needs children. It is no longer sufficient to allow public schools to be the primary educators of students with developmental disabilities. Training is available through the Feuerstein Institute and Equipping Minds.
3. Lifetime learning is imperative. The brain continues to develop over an entire lifetime. It is important to continue to engage in stimulating learning activities during adulthood and old age.
4. Teachers should see each student with new eyes and as capable of learning. An optimistic attitude is essential. The former ideas of categorizing children into "bright" or "not so bright" must be changed. This will only happen when teachers begin to engage with children by mediating how to learn and how to think.

5. Stop focusing on a diagnosis or a "label" of Autism, Fetal Alcohol syndrome, learning disabled, Down syndrome, or intellectual disability. It simply does not make sense to follow a deterministic view of development in light of the findings in neuroplasticity.
6. Dynamic assessments should replace static assessments. All academic and intellectual testing should be done with care in administration and interpretation.
7. Research indicates the need for the training to understand the best way to include and teach individuals with disabilities, educating church leadership in disability theology and support, and congregations accepting that all people are created in the *imago dei* valued, and can contribute to a faith community (Ault, Collins, & Carter, 2013).
8. Churches can bring in a guest speaker for training workshops on understanding Autism, intellectual disabilities, ADHD, and other learning challenges for parents, children, teachers, and youth ministry; provide educational materials and resources for developing a special-needs ministry in the church; provide an Afterschool or Summer Program for children and adults to integrate spiritual formation and cognitive formation; and hire a educational consult to observe the current disability programs, students, and teachers to determine any learning needs, teaching strategies, and adaptations needed.

CONCLUSIONS

Research suggests that it is possible to significantly improve fluid intelligence in children with cognitive impairments, using a comprehensive cognitive development program such as the Feuerstein Instrumental Enrichment, Bright Start, and Equipping Minds Cognitive Development Curriculum based on mediated learning experience. If the brain is constantly changing, it is possible to develop the thinking skills and increase the cognitive abilities for all children. Advances in brain imaging techniques allow us to understand and identify the cognitive neural systems to be strengthened. Neuroscience techniques provide valuable information for cognitive modifiability and hope for learners of all ages and etiologies (Tan & Seng, 2008).

Many Christian educators embrace the words of the great theologian and father of modern education, John Amos Comenius, who stressed the need to educate the intellectually and physically handicapped. According to Daniel Murphy, Comenius pleaded for educators to respond to those with special needs with extra sensitivity (Murphy, 1995). He believed that all humans are created in the image of God and have the capacity to learn:

> It is evident that man is naturally capable of acquiring knowledge of all things since, in the first place, he is the image of God. So unlimited is the capacity of the mind that in the process of perception, it resembles an abyss ... for the mind, neither in heaven nor anywhere outside heaven, can a boundary be fixed. The means to wisdom are granted to all men, and he reaffirms the common character of learning potentiality in all of mankind. What one human being is or has or wishes or knows or is capable of doing, all others are or have or wish or know or are capable likewise. (pp. 87–89)

Let us join Feuerstein and Comenius by embracing a belief in modifiability and give our children the opportunity to reach all that God created them to be.

DISCUSSION QUESTIONS

1. What is intelligence, and how can it increase? Explain based on Reuven Feuerstein's theory of Structural Cognitive Modifiability.
2. Many psychologists and educators continue to hold to a fixed view of intelligence for students with special needs. Agree or disagree? Explain.
3. Can a Christian school meet the needs of children with special needs such as Down syndrome, learning disabilities, and Autism Spectrum Disorder?
4. Since we know that cognitive abilities are modifiable, what should be the response of the Christian educator? Should we allow the public schools to be the primary educators of special needs students?
5. How does the Bible define the role of a mediator? What are the three steps for mediation? Is a human mediator superior to a computer program for learning?
6. Do you believe conventional intelligence tests are accurate indicators of a person's abilities? How would Reuven Feuerstein respond to that question?
7. What value do you place on intelligence tests? Should dynamic assessments replace them?

REFERENCES

Alloway, T. (2011). *Improving working memory: Supporting students' learning*. London, England: SAGE.

Alloway, T., & Alloway, R. (2014). *The working memory advantage: Train your brain to function stronger, smarter, faster.* New York, NY: Simon & Schuster.

Ault, M. J., Collins, B., & Carter, E. W. (2013). Congregational participation and supports for children and adults with disabilities: Parent perceptions. *Intellectual and Developmental Disabilities, 51*(1), 48–61.

Bohács, K. (2014). *Clinical applications of the modifiability model: Feuerstein's mediated learning experience and the instrumental enrichment program* (Doctoral dissertation). Graduate School of Educational Sciences, University of Szeged, Hungary.

Boleyn-Fitzgerald, M. (2010). *Pictures of the mind: What the new neuroscience tells us about who we are.* Upper Saddle River, NJ: Pearson Education.

Boyd, B. (2007). Intelligence is not fixed. *Journey to excellence.* Retrieved from http://www.journeytoexcellence.org.uk/videos/expertspeakers/intelligenceisnotfixedbrianboyd.asp

Camos, V. (2008). Low working memory capacity impedes both efficiency and learning of number transcoding in children. *Journal of Experimental Child Psychology, 99*, 37–57.

Doidge, N. (2015). *The brain's way of healing.* New York, NY: Viking.

Feuerstein, R., & Lewin-Benham, A. (2012). *What learning looks like: Mediated learning in theory and practice, K–6.* New York, NY: Teachers College Press.

Feuerstein, R., Feuerstein, R. S., & Falik, L. H. (2010). *Beyond smarter: Mediated learning and the brain's capacity for change.* New York, NY: Teachers College Press.

Feuerstein, R., Falik, L. H., & Feuerstein, R. S. (2015). *Changing minds & brains.* New York, NY: Teachers College Press.

Feuerstein, R., & Rand, Y. (1997). *Don't accept me as I am.* Arlington Heights, IL: Skylight.

Feuerstein, R., Feuerstein, R. S., Falik, L. H., & Rand, Y. (2006). *The Feuerstein instrumental enrichment program.* Jerusalem, IL: ICELP Publications.

Haywood, H. C. (2004). Thinking in, around, and about the curriculum: The role of cognitive education. *International Journal of Disability, Development, and Education, 51*(3), 231–252.

Katims, D. S. (2000). The quest for literacy: Curriculum and instructional procedures for teaching reading and writing to students with mental retardation and developmental disabilities. *Mental Retardation and Developmental Disabilities, Prism Series, 2*, 3–14.

Klauer, K. J. (2002). A new generation of cognitive training for children: A European perspective. In G.M van der Aalsvoort, W. Resing, & A. Ruijssenaars (Eds.), *Learning potential assessment and cognitive training* (pp. 147–174). New York, NY: Elsevier Science.

Kozulin A., Lebeer, J., Madella-Noja, A., Gonzalez, F., Jeffrey, I., Rosenthal, N., & Koslowsky, M. (2010). Cognitive modifiability of children with developmental disabilities: A multicenter study using Feuerstein's instrumental enrichment–basic program. *Research in Developmental Disabilities, 31*(2), 551–559.

Lane, J., & Kinnison, Q. (2014). *Welcoming children with special needs.* Bloomington, IN: West Bow Press.

Mentis, M., Dunn-Bernstein, M., Mentis, M., & Skuy, M. (2009). *Bridging learning: Unlocking cognitive potential in and out of the classroom.* Thousand Oaks, CA: Corwin.

Murphy, D. (1995). *Comenius: A critical reassessment of his life and work.* Portland, OR: Irish Academic Press.

Nevin, A. (2000). *Lesson plans for self-determination across the K–12 curriculum for students with learning disabilities, students with mental retardation, students with emotional disabilities, and students with traumatic brain injury.* Phoenix, AZ: Arizona State University.

Ormrod, J. (2010). *Educational psychology: Developing learners* (7th ed.). Upper Saddle River, NJ: Merrill Prentice-Hall.

Paour, L. (1993). Induction of logic structures in the mentally retarded. In C. Haywood & D. Tzuriel (Eds.), *Interactive assessment* (pp. 119–166). New York, NY: Springer.

Patel, V., Aronson, L., & Divan, G. (2013). *The school counsellor casebook.* Manipal, IN: Byword Books.

Piaget, J. (1973). *The child and reality: Problems of genetic psychology.* New York, NY: Grossman.

Piaget, J. (1952). *The origins of intelligence in children.* New York, NY: International Universities Press.

Sternberg, R. (1984). How Can We Teach Intelligence? *Educational Leadership*, 9.

Tan, O. S., & Seng, S. H. A. (2008). *Cognitive modifiability in learning and assessment: International perspectives.* Singapore: Cengage Learning.

SUGGESTIONS FOR FURTHER READING

Alloway, T., & Alloway. R. (2014). *The working memory advantage: Train your brain to function stronger, smarter, faster.* New York, NY: Simon & Schuster.

Feuerstein, R., & Lewin-Benham, A. (2012). *What learning looks like: Mediated learning in theory and practice, K–6.* New York, NY: Teachers College Press.

Feuerstein, R., Falik, L. H., & Feuerstein, R. S. (2015). *Changing minds & brains.* New York, NY: Teachers College Press.

Mentis, M., Dunn-Bernstein, M., Mentis, M., & Skuy, M. (2009). *Bridging learning: Unlocking cognitive potential in and out of the classroom.* Thousand Oaks, CA: Corwin.

Tan, O. S., & Seng, S. H. A. (2008). *Cognitive modifiability in learning and assessment: International perspectives.* Singapore: Cengage Learning.

ABOUT THE EDITORS

Mark A. Maddix, PhD, is a professor of practical theology and Christian discipleship and the dean of the School of Theology and Christian Ministries at Point Loma Nazarene University, San Diego, California. Reverend Maddix served in pastoral ministries for 12 years and in teaching ministry for 18 years. He has served as the president of the Society of Professors in Christian Education. He also has coauthored and coedited the following books: *Discovering Discipleship: Dynamics of Christian Education* (2010), *Spiritual Formation: A Wesleyan Paradigm* (2011), *Best Practices of Online Education: A Guide to Christian Higher Education* (2012); *Missional Discipleship: Partners in God's Redemptive Mission* (2013); *Pastoral Practices: A Wesleyan Paradigm* (2013); *Essential Church: A Wesleyan Ecclesiology* (2014). He has published more than 40 articles of book chapters in the field of Christian education, spiritual formation, and online education. Mark and his wife Sherri have two grown children, Adrienne Meier and Nathaniel Maddix.

Dean G. Blevins, PhD, is a professor of practical theology and Christian discipleship at Nazarene Theological Seminary, Kansas City, Missouri. An active scholar, Reverend Blevins has contributed to several books and published more than 70 church-related or scholarly articles. In addition, Dean cowrote the text *Discovering Discipleship: The Dynamics of Christian Education,* and currently serves both as an editor of the *Horizons in Religious Education* book series, and as senior editor of *Didache: Faithful Teaching,* an online academic journal. Dr. Blevins also served as president of the Religious Education Association and currently serves several editorial boards including *Theological Education* and the *Journal of Family & Community Ministries*. He consults regularly in children, youth, and family ministry, as well as guides clergy development in the United States and Canada for the Church of the Nazarene. Dean also served as board chairman of YouthFront, a Kansas City-based national youth ministry training organization. Dean lives in Olathe Kansas with his wife JoAnn and daughter Rachel.

BOOK CONTRIBUTORS

Glena Andrews, PhD, is a professor of clinical psychology and director of clinical training at George Fox University, Newberg, Oregon.

Laura Barwegen, EdD, is an associate professor of Christian formation and ministry at Wheaton College, Wheaton, Illinois.

Carol T. Brown, EdD (candidate), is the executive director and an educational specialist at Equipping Minds in Danville, Kentucky.

Warren Brown, PhD, is a professor of psychology in the Graduate School of Psychology and director of the Lee Edward Travis Research Institute, Fuller Theological Seminary, Pasadena, California.

Karen Choi, PhD, is a professor of spirituality and Christian education and director of international student affairs and Presbyterian Theological Seminary in America, Santa Fe Springs, California.

James R. Estep, PhD, is a professor of Christian education at Lincoln Christian University, Lincoln, Illinois.

Theresa A. O'Keefe, PhD, is a professor of practice and codirector of Contextual Education at Boston College, School of Theology, Boston, Massachusetts.

Brad Strawn, PhD, is the Evelyn and Frank Freed Professor of the Integration of Psychology and Theology at Fuller Theological Seminary, Pasadena, California.

John David Trentham, PhD, is an assistant professor of leadership and discipleship and director of the Doctor of Education Program at Southern Baptist Theological Seminary, Louisville, Kentucky.

Timothy Paul Westbrook, PhD, is an assistant professor of Bible and director of the Center of Distance Education in Bible and Ministry at Harding University, Searcy, Arkansas.

www.ingramcontent.com/pod-product-compliance
Lightning Source LLC
Chambersburg PA
CBHW070616300426
44113CB00010B/1545